This Book Belongs to:

....

....

Wild
IN THE
CITY

A GUIDE TO URBAN ANIMALS
AROUND THE WORLD

Written by Kate Baker | Illustrated by Gianluca Folì

CONTENTS

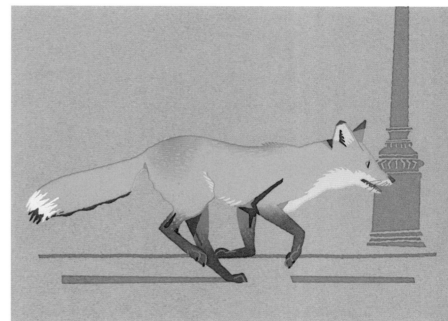

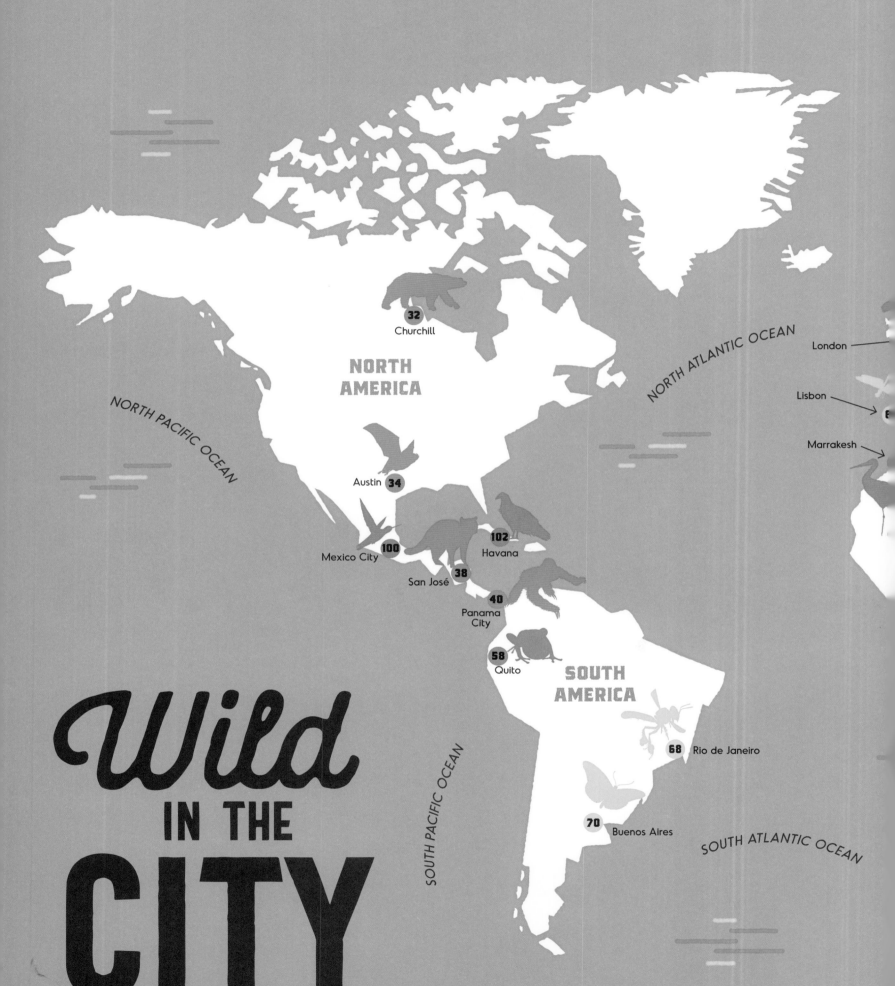

32
Churchill

NORTH AMERICA

NORTH PACIFIC OCEAN

NORTH ATLANTIC OCEAN

London

Lisbon

Marrakesh

Austin **34**

Mexico City **100**

San José **38**

Havana **102**

Panama City **40**

58 Quito

SOUTH AMERICA

SOUTH PACIFIC OCEAN

68 Rio de Janeiro

70 Buenos Aires

SOUTH ATLANTIC OCEAN

Wild
IN THE
CITY

ARCTIC OCEAN

ASIA

86 Helsinki

42 St Petersburg

84 Amsterdam

EUROPE

celona

82

Naples

16 Cairo

Beijing 94

26 Nara

NORTH PACIFIC OCEAN

Dakha 22

Mumbai 20

18 Harar

Bangkok 52

56 Manila

AFRICA

28

Singapore

54 Jakarta

50 Antananarivo

AUSTRALASIA

INDIAN OCEAN

Sydney 96

66 Wellington

92 Simon's Town

Mammals   Reptiles / Amphibians   Bugs   Birds

The Urban Jungle

CRAMMED WITH BUILDINGS, TRAFFIC AND PEOPLE, URBAN SPACES ARE THE LAST PLACE YOU WOULD EXPECT TO SEE WILDLIFE. But all kinds of animals live alongside us in the hidden corners of our towns and cities — from teeny ants living under pavement cracks to pick-pocketing monkeys and bin-raiding polar bears!

Some choose to move in — cities are like fully stocked larders and there are lots of cosy nooks where animals can sleep and raise their families. Others lived there long before the city was built and people moved in, and have learned to adapt to their new surroundings.

Within these pages you'll travel from city to city, across six different continents, meeting just some of these amazing animals. There are tips on where and when you might see them, what signs to look for, and how you can help make our cities more nature-friendly places. Many animals are either shy or speedy, but slow down, take a closer look around you and you might even experience an incredible wildlife encounter of your own!

CONSERVATION STATUS

As our cities continue to grow and wild habitats are lost, some species are in danger of being lost forever. On each page you will see the conservation status of the animal featured. For this we have used the IUCN Red List of Threatened Species grading from Least Concern to Near Threatened, Vulnerable, Endangered, Critically Endangered, Extinct in the Wild, and finally, Extinct.

MAMMALS

MAMMALS IN THE CITY

SOME ANIMALS CAN BE HARD TO SEE. They might only come out at night, or they might be very good at hiding. But with a bit of practice, you can learn to spot some of the clues that they leave behind.

Tracks and Trails

As animals move across the ground they leave behind footprints — known as tracks. By looking at these marks you can work out what animal they belong to and even what they've been up to. The best place to look for animal tracks is on mud or sand after a spot of rain, or on freshly fallen snow.

Paw or hoof?

WHAT TO LOOK FOR

First of all, see if you can tell what kind of footprint it is. Is it a paw or is it a hoof?

Paw
(e.g. cats, dogs, rabbits, bears, racoons, weasels and rodents)

Hoof
(e.g. deer, pigs, sheep, cows and goats)

A closer look

Now take a closer look. What is the overall shape? How big is it? How many digits (toes) are there? What about the claws? Can you see them?

Keep an eye out in city parks or in your garden!

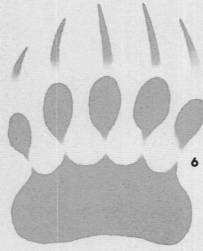

1. Human
Our footprints look like this!

2. Fox
Fox paws are sort of diamond shaped. Sometimes claw marks are visible.

3. Squirrel
Squirrels have long claws to help them grip onto trees.

4. Otter
Otter prints have webbing between the toes.

5. Deer
Deer don't have paws, they have hooves. Each hoof is divided in two.

6. Bear
Bears put their whole feet on the ground, just like us.

7. Leopard
Most cats have retractable claws, so they don't leave claw marks.

Fox or dog?

Next, see how many prints there are and how they are arranged. This can tell you how the animal walks or runs.

1. Foxes trot along in a nice neat line. They are often on a mission to find food.
2. Dog tracks are a bit more chaotic, like this.

Whose Poo?

Droppings are another sign that an animal has been near. Just like the animals themselves, animal poo comes in different shapes and sizes. Some is round, some long and thin, some is hard, some sloppy. Some contains feathers or bones and can tell you what the animal had for dinner.

1. Red squirrel
Keep your eyes peeled for round, oval or pellet-shaped droppings.

2. Bat
Bat droppings are brown or black and shaped like tiny pellets.

3. Red fox
Fox poo is long and thin, with a pointy bit at one end. It can be different colours depending on what the fox has eaten.

4. Deer
Deer poo are round pellets. Sometimes they are stuck together in clumps.

5. Otter
Otter poo smells a bit fishy and usually contains fish bones, scales or feathers. They can be found perched on top of rocks and on riverbanks.

6. Hedgehog
Hedgehogs do surprisingly big poos! They sometimes contain beetles and worms.

7. Wild boar
Boar poo is usually sausage shaped, or like big round dumplings.

REMEMBER

Don't touch the poo you find, and watch where you step!

Squeeeeeeeak squeeeeeeeeak!

Squeeeeeeeak!

OTHER CLUES

As well as prints, tracks and droppings, there are other things you can look out for, too…

- Bits of fur snagged on bushes
- Nibble marks on nuts and fruit
- Mounds of earth kicked up by burrowing animals such as moles
- Claw marks on trees from hungry bears!

GRUNTS AND GROWLS

You might not be able to see an animal, but you may be able to hear it. The best time is at night, when the city goes to sleep. Listen out for grunts, growls, squeaks and howls.

RED FOXES
in London, United Kingdom

LONDON IS FAMED FOR ITS BRIGHT RED BUSES, HISTORIC BUILDINGS AND BUCKINGHAM PALACE, HOME OF THE QUEEN. The city is also home to around 10,000 red foxes. These wily, bushy-tailed mammals can often be seen trotting down leafy streets, rifling through bins or tiptoeing along the tops of garden fences. One was even spotted riding a London Underground train!

Triangular ears that point forwards to help it listen out for prey in the darkness

Long whiskers that help it sense the movement of nearby creatures

LATIN NAME *Vulpes vulpes*
FAMILY Canidae (mammal family that includes foxes, dogs, wolves and coyotes)
LENGTH 45—90 cm (1.5—3 ft) plus 30—55 cm (12—25 in) tail
CONSERVATION STATUS
Least Concern
WHERE IN THE WORLD? Native to North America, Europe, Asia, and parts of North Africa; introduced to Australia

Black legs that look a bit like socks

Foxes have excellent night vision and amazing hearing. They can hear a mouse squeak over 30 m (98 ft) away!

DID YOU KNOW?

Male foxes are called dogs. Females are called vixens and baby foxes are called cubs, pups or kits. A group of foxes is known as a skulk.

Thick fur coat

Sharp teeth to catch and chew its prey

Magnificent bushy tail with a white tip

Foxes have a very good sense of smell — they can smell small creatures up to 1 m (3.2 ft) underground.

A fox's tail is known as a brush. It helps keep the fox steady as it runs, jumps or climbs.

City Life

Red foxes are a common sight in many cities around the world, especially around the edges (or suburbs). The suburbs are a bit like a bridge between the town and the countryside. In London, there are lots of houses with gardens and allotments with sheds, which are great places for making dens and sniffing out food. Some people think foxes are a bit of a nuisance, especially when they rummage through bins or pull up flowers while looking for slugs. But they do a very good job of gobbling up the city's rats and pigeons. And remember, the city is *their* home, too. If they find food lying about, can you blame them for tucking in?

UNDERGROUND HOMES

In towns and cities, foxes build their dens underneath sheds, hedges or bushes, in railway embankments, or under piles of wood. Their underground homes are a warm, safe place for the cubs to be born, protected from predators.

A fox's home is called a den.

SPOTTED

Foxes have been seen in some surprising places all over London ...

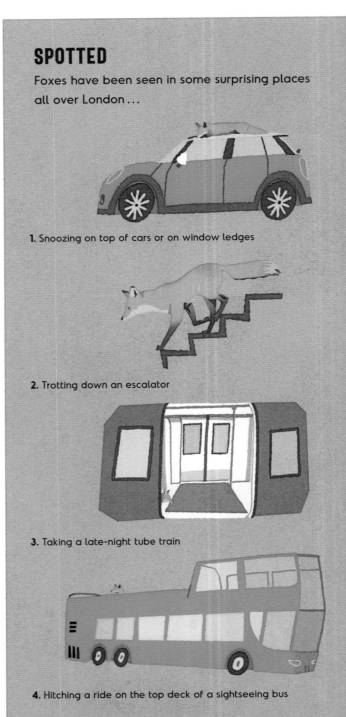

1. Snoozing on top of cars or on window ledges

2. Trotting down an escalator

3. Taking a late-night tube train

4. Hitching a ride on the top deck of a sightseeing bus

CITY SLICKER

One London fox made its home at the top of the Shard building — a huge skyscraper made of glass. To reach the top, this intrepid fox, nicknamed 'Romeo', had to climb 71 sets of stairs and a wooden ladder!

GETTING ABOUT

Foxes are very quick and agile. They can jump over high fences and squeeze through tight spaces. This helps them to move easily around towns and cities.

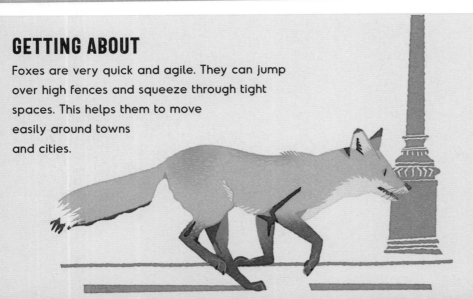

WHAT DO THEY EAT?

Foxes are omnivores, so they're just as happy munching on berries and fruit as on small mammals, birds and frogs. One of their favourite foods is earthworms — sometimes they eat hundreds in one night! There are also easy meals to be found around the backs of cafes and restaurants and in rubbish bins.

COMMUTERS

Just like people, some foxes 'commute', travelling into the city from the countryside at night to find food. Others may start their lives in the city then move to the countryside as they get older.

Spotting Foxes

WHERE TO SEE THEM

Close to their dens and wherever there's food to be found. In gardens or allotments digging up slugs and worms or hunting mice. Hanging about near houses and shops looking for scraps. In cemeteries or along railway tracks.

WHEN TO SEE THEM

Foxes are mainly active at night, but they can sometimes be seen in the daytime, too. The best time to look for them is before you go to bed, or early in the morning before the sun comes up. In springtime, watch out for young cubs play-fighting in the grass near their den.

LISTEN OUT FOR ...

'A-woo' — a barking noise, a bit like that of a domestic dog, made by male foxes
'Eeeeee' — a spine-chilling scream made by female foxes when calling out for a mate
'Ack-ack-ack-ack' — the noise made by young cubs as they play-fight

EGYPTIAN WEASELS
in Cairo, Egypt

On the edge of the city are the amazing Pyramids of Giza — the last surviving Wonder of the Ancient World.

The Museum of Cairo is filled with treasures discovered in the tombs of the ancient pharaohs.

Weasels have lived close to humans for thousands of years. In Ancient Egypt they were kept as pets, alongside cats, dogs, falcons, gazelles, baboons, lions and hippos!

Small head and long thin body that allows it to squeeze into narrow spaces

The weasel uses its amazing sense of smell to sniff out food in the darkness.

Weasels have excellent hearing and night vision, helping them to hunt day or night.

RISING FROM THE DESERT, CAIRO IS ONE OF AFRICA'S BIGGEST AND BEST-KNOWN CITIES. In the daytime, cars and trucks share the dusty roads with donkeys, camels, chickens and ducks. At night, the backstreets are the hunting ground for one of the smallest carnivores on the planet — the Egyptian weasel. Famed for their amazing ability to fit through openings as small as a ping pong ball, these nifty mammals are pretty good at staying out of sight, but if you're lucky, you might glimpse one slinking under a parked car or disappearing into a pavement crack.

LATIN NAME *Mustela subpalmata*
FAMILY Mustelidae (a mammal family that includes weasels, otters, badgers, ferrets, martens and wolverines)
LENGTH 32–43 cm (12–17 in) plus 9–13 cm (3.5–5 in) tail
CONSERVATION STATUS Least Concern
WHERE IN THE WORLD? Egypt

City Life

These adaptable little animals don't need green spaces or trees to feel at home in the city. They mainly live in buildings, under piles of rocks, in water pipes or inside wall crevices. And they're not fussy about what they eat. They'll devour pretty much whatever they can get their claws on — from fruit and vegetables to mice, rats, lizards, birds, rabbits, fish, chicken eggs and cockroaches.

Cairo's weasels sneak into restaurants, raid food markets and dig around rubbish bins for scraps.

GETTING ABOUT

Thanks to their long, slender bodies, weasels can get into nooks and crannies that larger predators can't. They can even fit through pavement cracks and reach rodents nesting in underground burrows.

FEARSOME HUNTERS

Don't let their cute faces deceive you! The fearless weasel is as ferocious as a lion and is not afraid to attack animals more than twice its size. First it kills its victim with a bite to the base of its neck, then drags the carcass back to its den. In Shanghai, China, yellow weasels are caught in cages and then set free in rat-infested parts of town to help clear the streets.

One hungry weasel was spotted clinging onto the back of a woodpecker in a park in London, UK.

WEASEL TALES

In Cairo, some people believe that weasels steal gold or shiny objects. If a ring or earring goes missing, the weasel often gets the blame. In China, some people say that seeing a weasel is good luck. Others believe these crafty creatures can steal people's souls.

GOOD NEIGHBOURS

Weasels get along with their human neighbours pretty well most of the time. But sometimes they move into the roof spaces of houses while people are still living there or steal eggs from chicken coops.

Spotting Weasels

WHERE TO SEE THEM

Peeping their head out of cracks and crevices. Or making a run for it across the street then hiding underneath parked cars.

WHEN TO SEE THEM

They can sometimes be seen in the day, but they usually come out at night.

LISTEN OUT FOR...

Yips, cries, snorts and hisses. Or the pitter-patter of footsteps as they run across rooftops in the middle of the night.

SPOTTED HYENAS
in Harar, Ethiopia

Hundreds of mosques, shrines and tombs are dotted around the city.

Strong jaws and sharp teeth for tearing through flesh and eating bones

Massive neck and large head

LATIN NAME
Crocuta crocuta
FAMILY Hyaenidae (hyenas)
LENGTH 0.95–1.85 m
(3–6 ft) plus 25–30 cm
(10–12 in) tail
CONSERVATION STATUS
Least Concern
WHERE IN THE WORLD?
Sub-Saharan Africa

City Life

So the story goes, there was once a great famine in Harar and the starving hyenas began to attack humans and livestock. Local leaders decided to leave food out for the hyenas to stop them from killing people and animals. The custom continues to this day. Keeping the hyenas fed means they're less likely to go after the town's goats, sheep and cows. The local hyenas also protect the city from packs of hyenas from other parts of the countryside.

The hyenas help to keep the city clean by eating food waste from the meat markets and rubbish dumps.

BONE CRUNCHERS

On the African grasslands, spotted hyenas hunt in packs to chase down zebras, warthogs and gazelles. They also scavenge on kills left behind by lions and other predators. In Harar, local butchers leave out scraps of meat and leftover bones, and the hyenas crunch through them with their powerful jaws and sharp teeth.

AS NIGHT FALLS OVER THE WALLED CITY OF HARAR, packs of wild spotted hyenas leave their nearby caves and creep into the town in search of food. In other African cities these fearsome predators have been known to kill livestock or attack humans, but the people of Harar are not afraid. They have been welcoming hyenas for hundreds of years, leaving out scraps of food for them to eat and even feeding them by hand!

Yellowish or grey coat with dark spots

Small doorways were built into the city walls to allow the hyenas to come and go more easily.

Hyenas have one of the strongest bites of any mammal.

'HYENA MEN' OF HARAR

The 'hyena men' of Harar feed the hyenas with their hands or from sticks in their mouths. They believe that by eating the meat the hyenas are also eating the evil spirits that haunt the town.

PREDICTING THE FUTURE

During the Muslim festival of Ashura, the hyenas are summoned by drums to shrines, where they are fed a special porridge prepared by Harar's people. According to tradition, if the hyenas eat more than half of the porridge, the city will have a peaceful and prosperous year. If they refuse to eat, there will be famine or disease.

Spotting Hyenas

WHERE TO SEE THEM

Roaming near rubbish bins or meat markets.

WHEN TO SEE THEM

After dark. They rest in nearby caves during the day but enter the city at night to look for food.

LISTEN OUT FOR ...

'Hee-hee-hee' — the eerie high-pitched cackle of the hyena. It sounds a lot like hysterical human laughter, but this call is their way of letting other hyenas know that a member of the pack has made a kill or been attacked, or that danger is near.

By day, the city is buzzing with people and traffic. At night, the streets are hunting grounds for leopards.

LATIN NAME *Panthera pardus*
FAMILY Felidae (cats)
LENGTH 1.3–1.9 m (4.2–6.2 ft) plus 1–1.4 m (3.5–4.5 ft) tail
CONSERVATION STATUS Vulnerable
WHERE IN THE WORLD? Parts of Africa and Asia

A leopard's spots are called 'rosettes' because they look a bit like roses. They help the leopards blend in with their surroundings.

Excellent hearing and night vision so it can hunt in the dark

Sharp, curved retractable claws

Powerful jaw muscles for holding onto its prey

LEOPARDS
in Mumbai, India

WITH A POPULATION OF AROUND 20 MILLION PEOPLE, MUMBAI IS ONE OF THE BIGGEST CITIES ON THE PLANET. But in the midst of this noisy concrete jungle is a surprising haven for wildlife — the Sanjay Gandhi National Park. This huge tropical forest is teeming with animals and plants — there are butterflies, birds, monkeys, hyenas, porcupines, hares, deer, crocodiles, cobras . . . and around 40 wild leopards!

The Sanjay Gandhi National Park has one of the highest densities of leopards anywhere on Earth.

City Life

Most of Mumbai's leopards live inside the National Park. But leopards don't know the difference between a 'wild' habitat and a human one. Under the cover of darkness, they venture out of the forest to prowl the nearby neighbourhoods for food. The idea of leopards stalking the city streets may sound scary, but attacks on people are rare. Their main targets are stray dogs and livestock such as pigs, chickens and goats.

Leopards lived in the park's forests long before the city was built.

SPOTTED!

One leopard was caught on security camera creeping into an apartment block and leaving a little later with a small dog in its jaws. Another was seen paying a visit to the five-star Renaissance Hotel!

THE DOGS OF MUMBAI

Mumbai is also home to thousands of stray dogs. They hang around street corners and alleyways, attracted to the rubbish that is dumped there. These dogs are much easier for leopards to hunt than the wild deer that run around inside the park.

People throw food into the street and it attracts stray dogs and pigs.

Spotting Leopards

WHERE TO SEE THEM

On the edges of cities near their forest homes. Stalking prey near rubbish dumps.

WHEN TO SEE THEM

The best chance is at night, when they come out to hunt. In the day, leopards are usually well hidden as they snooze in trees or shelter in shady caves.

LISTEN OUT FOR ...

Leopards are silent a lot of the time, but when they're cross they may growl, roar or spit. When they're content they purr, just like domestic cats!

LOOK OUT FOR ...

'Eye shine' — when the leopards' eyes glow like lights in the darkness.

SILENT HUNTERS

Leopards prefer to creep up on their prey rather than chase after it. They stalk their victims quietly and slowly, with their head bent low so they can't be seen. When they're only a few metres away, they pounce!

The noise of the city makes it easier for the leopards to sneak up unheard.

BIG CATS IN THE CITY

Mumbai is not the only city with big cats. In Los Angeles, USA, the home of movie stars, a huge mountain lion named P-22 roams the Hollywood Hills.

RHESUS MACAQUES
in Dhaka, Bangladesh

Dhaka is the capital of Bangladesh and its biggest city.

THE CITY OF DHAKA IS BURSTING WITH PEOPLE AND BUILDINGS. It is also home to hundreds of mischievous macaque monkeys that patrol the streets in noisy troops. They have settled into this chaotic and crowded habitat amazingly well, leaping from the rooftops, raiding food stalls and grabbing food from passers-by. Some even sneak into people's houses!

Long furry tail for balancing and leaping

Reddish-pink faces (and bottoms)

Expressive faces

Forest trees have been swapped for buildings and cars.

LATIN NAME *Macaca mulatta*
FAMILY Cercopithecidae (Old World monkeys)
LENGTH 45–64 cm (1.5–2 ft) plus 19–32 cm (3.5–4.5 ft) tail
CONSERVATION STATUS Least Concern
WHERE IN THE WORLD? Parts of Asia

The macaques live in big noisy troops of up to 200 animals.

Large cheek pouches for storing food to eat later

City Life

There used to be many types of monkey living in Dhaka, but as the city grew and there were fewer trees and green spaces, many of them moved out. Rhesus macaques don't mind busy places. They are expert leapers and climbers and get around town a bit like they do in their jungle habitat — jumping from roof to roof, sliding down telegraph poles and swinging from washing lines!

1. The buildings in old Dhaka are very close to one another, making it easy for the macaques to leap from place to place.

2. Rooftop water tanks are the perfect place to cool off!

3. Electrical wires remind the monkeys of jungle vines. They use them to cross over busy roads, but this can be very dangerous. Many monkeys have been electrocuted and killed while running along the wires. Others have fallen injured onto roads and train tracks.

PLAY TIME

There's plenty of fun to be had in the city too! Macaques are good swimmers and love dive-bombing into Mumbai's fountains and pools.

MONKEY FACTORY

One troop of monkeys has set up home in an old factory in Dhaka's backstreets. Next to the factory is a graveyard with green space for them to play in, and every day they are fed lentils or chickpeas by the monkey-loving factory owners.

MONKEYING AROUND

Some people think the monkeys are like hooligans — breaking into houses, pinching food and making lots of noise and mess as they zoom around the neighbourhoods. Others see the monkeys as part of the community and enjoy feeding them or watching them play.

URBAN MONKEYS OF THE WORLD

In Jaipur, India, a troop of langur monkeys lives in a temple dedicated to the Hindu monkey god, Hanuman. The monkeys are treated like royalty and pilgrims feed them every day.

At the Buddhist temple of Swayambhunath in Kathmandu, Nepal, rhesus macaques swing from the colourful prayer flags and slide down the railings.

At the Uluwatu Temple in Bali, Indonesia, crab-eating macaques steal people's purses, hats or sunglasses — then return them to the owner in exchange for food!

The feisty chacma baboons of Cape Town, South Africa, have learnt to break into cars. They jump in, grab what food they can, then make a run for it.

Spotting Monkeys

WHERE TO SEE THEM

They thrive in many big cities around the world. In Asia, they often live near Hindu communities, shrines and temples. Spot them climbing on buildings, crossing roads, hanging around markets, on railway platforms, or breaking into houses and cars.

WHEN TO SEE THEM

Anytime! They are active in the day and at night.

LISTEN OUT FOR . . .

Monkeys use noises and facial expressions to talk to one another. When they smile it sometimes means they're scared and don't want you to come any closer.

TREE-LOVING TAMARINS

Not all monkeys are adapting to life in the city. In Manaus, Brazil, the tiny pied tamarin prefers to live in the treetops, but the city is growing, leaving only small patches of forest behind. These tamarins are now the most threatened primates in the Amazon and are in danger of extinction.

25

SIKA DEER
in Nara, Japan

DEER ARE USUALLY SHY CREATURES AND WILL RUN AWAY WHEN PEOPLE GET CLOSE TO THEM. But in the Japanese city of Nara, they have been living side by side with humans for centuries. According to local folklore, long ago a god rode into Nara on the back of a white deer. Since then, they have been thought of as heavenly animals and left to freely roam the streets and parks.

Male deer are called stags. They use their antlers to impress the ladies and defend their territory.

Baby deer are called fawns.

Females are called hinds or cows.

Pale bottom and spots their back

City Life

LATIN NAME *Cervus nippon*
FAMILY Cervidae (deer)
LENGTH 0.9–1.8 m (3–5.9 ft) plus 7.5–13 cm (3–5 in) tail
CONSERVATION STATUS Least Concern
WHERE IN THE WORLD? Native to China, Japan and Russia; introduced to parts of Europe, USA and New Zealand

Sika deer have been living in Nara since the city was founded over 1000 years ago. They behave much the same way in the town as they do in the forest. They wander around grazing on plants, eating grass in Nara Park and nibbling on weeds from the roadside. They feel so at home here that they've even been seen walking into temples and down the aisles of supermarkets!

As day breaks, hundreds of deer leave the nearby forest and make their way into town. They've been spotted in some unusual places!

Waiting for a bus

Crossing the road

Going to the shops

Having a snack

Nara Park is the best place to spot the deer — more than 1200 live here.

There are also many ancient temples and shrines inside the park.

Some deer have learned to bow, which is a sign of respect in Japan.

Vendors sell special rice and wheat crackers called *shika senbei* so visitors can feed the deer.

鹿せんべい 150円

NATIONAL TREASURES

The sika deer are a symbol of the city and are carefully protected. Park signs warn visitors not to harm the deer.

1. **No Littering** — deer can eat the rubbish and get sick

2. **No Feeding** — except for shika senbei deer crackers as they are made especially for deer

3. **Keep Dogs on Lead**

4. **Do Not Hit or Chase the Deer**

Spotting Deer

WHERE TO SEE THEM

In city suburbs close to forests and parks. In Nara, they can be seen all around the city and its parks and temples.

WHEN TO SEE THEM

Usually, deer are most active from early evening to early morning. In Nara they can also be seen in the daytime.

LISTEN OUT FOR...

Deer communicate using all kinds of sounds from soft whistles to loud screams. Stags sometimes blow raspberries to impress the female deer!

LOOK OUT FOR...

Scratch marks on tree bark where male deer have rubbed their antlers against the trees.

SMOOTH-COATED OTTERS
in Singapore

THE ISLAND OF SINGAPORE IS RENOWNED FOR ITS SOARING SKYLINES, SHINY SHOPPING MALLS and squeaky-clean streets. It is also one of the world's greenest cities. It has its own botanic gardens crammed with plants and flowers, a rainforest nature reserve, a zoo and a bird park with around 400 species of bird. Its most famous wild animal residents are the families of smooth-coated otters that dine on fish and frolic in the waterways just a few minutes from the city centre.

Keep your eyes peeled for their round brown heads bobbing above the water!

Long whiskers that detect the vibration of prey underwater

If they hear a strange noise, otters lift their heads out of the water like a periscope and have a good look around.

After dinner, the otters head for land to dry off. They roll around to rearrange their fur then mark their territory by leaving droppings behind.

Webbed feet

Long flat tail that helps it to steer and swim faster

LATIN NAME *Lutrogale perspicillata*
FAMILY Mustelidae (mammal family that includes otters, wolverines, weasels, badgers, ferrets and martens)
LENGTH 59—64 cm (1.9—2 ft) plus 37—43 cm (1.2—1.4 ft) tail
CONSERVATION STATUS Vulnerable
WHERE IN THE WORLD? Southern Asia and a small area of Iraq

One otter family has taken up residence in the world-famous Marina Bay, right in front of the luxury Marina Bay Sands Hotel.

Thick short brown fur to keep it warm

Smooth, sleek coat and streamlined body

City Life

In the 1970s, otters were thought to have disappeared from Singapore. Years of river traffic and industries on the riverbank had left the water filled with rubbish and black with pollution. Since then, the waterways have been cleaned up and the otters have returned. Today, there are around 70 otters living in the city, and they love putting on a show for passers-by!

OTTER WATCH

The otters have a big following among local Singaporeans and tourists. They even have their own Facebook page where people can record otter sightings and upload photo and videos.

EXPENSIVE FISHY SNACKS

Now the city's rivers and reservoirs are full of fish again, there's plenty for the otters to eat. But they've also been known to sneak into exclusive hotels and private homes in the dead of night to feast on ornamental koi carp living in the fish ponds.

Otters were suspected of eating $80,000-worth of koi carp in one night!

ADAPTING TO THE CITY

Singapore's smooth-coated otters have adapted to city life in some surprising ways.

1. Some of Singapore's otters build their homes in holes dug into sandbanks or in the tall grass, just like they do in the wild. But this family is making do with a concrete canal.

2. Finding the best place to dry off isn't always easy. This adventurous otter was filmed using a ladder to climb out of a storm drain.

3. The sloping walls of the Marina Reservoir make it easy for the otters to climb out of the water.

PROTECTING OTTERS

Singaporeans like to look after their otters. Signs around the city tell you what to do if you spot one.

Watch out for otters crossing!

What to do if you encounter otters
- Do not touch, chase or corner the otters.
- Observe them from a distance.
- Keep your dog on a lead.
- Do not talk loudly or use flash photography.
- Do not feed the otters.
- Do not drop litter or sharp objects into the water.

CITY IN A GARDEN

Singapore is one of the greenest cities in the world. Its concrete skyline is slowly giving way to green skyscrapers, with lush rooftop gardens, and plants cascading down the walls. In Marina Bay, there is a grove of 50 m (164 ft) tall 'supertrees' — metal structures covered with thousands of plants.

SEA LIFE IN THE CITY

City waterways are home to other water-loving mammals, too. In San Francisco, USA, sea lions have been 'hauling out' at Fisherman's Wharf since 1989. It's a safe place for the sea lions to rest away from predators and is full of tasty herrings for them to eat.

People from all over the world come to see the famous sea lions on Pier 39.

Spotting Otters

Otters leave droppings behind to let other otters know where they live.

TOP TIP
Keep an eye out for their webbed tracks on muddy riverbanks.

Otters love to roll down the riverbanks on their bellies. Look out for their 'slide marks', which alternate with footprints.

WHERE TO SEE THEM

Otters have been making a comeback in towns and cities around the world. Look for them near rivers, reservoirs and lakes.

WHEN TO SEE THEM

They are most active first thing in the morning and early evening, when they like to hunt.

LISTEN OUT FOR ...

Otters are very chatty and talk to each other with clicks, whistles and chirps. When they feel threatened they hiss or growl.

POLAR BEARS
in Churchill, Canada

Churchill is known as the polar bear capital of the world.

A polar bear's nose is so powerful it can sniff out a seal up to 3 km (1.86 miles) away.

Some of the bears wander into town by mistake on their way to the sea. Others are attracted by the smell of food.

Thick white fur coat to keep it warm and hidden in the snow

Large paws for paddling through the water and moving across ice and snow

Giant curved claws to grab prey

Non-slip feet pads for gripping onto the ice

EVERY AUTUMN, HUNDREDS OF POLAR BEARS BEGIN THEIR GREAT JOURNEY NORTH to their hunting grounds on the Arctic ice caps. Directly in their path, perched on the shores of Hudson Bay, is the small town of Churchill. Many of the bears head straight for the beach but some sneak into the town itself, sniffing out scraps of food from the rubbish dump and scaring the residents into their houses!

LATIN NAME *Ursus maritimus* (meaning 'sea bear')
FAMILY Ursidae (bears)
LENGTH 1.8–2.6 m (5.9–8.5 ft)
CONSERVATION STATUS
Vulnerable
WHERE IN THE WORLD?
Arctic regions

City Life

Polar bears may look cuddly, but these mighty carnivores can kill most animals with one swipe of their giant paw. To keep the town safe, a team of rangers patrol the perimeters to scare away any bears that get too close. If that doesn't work, the bears are taken to the 'polar bear jail' then airlifted out of town and released back into the wild.

LIVING WITH BEARS

How to avoid a surprise encounter with a polar bear ...

1. Always check outside before leaving a building.
2. Keep to main streets and well-lit areas.
3. Don't go out alone after dark.
4. Be alert!

What to do if you meet a polar bear ...

1. Stay calm.
2. Back away, facing the polar bear at all times.
3. Get into a building or vehicle as quickly as possible.
4. Do not run and never chase a bear.

POLAR BEAR ALERT

STOP!

SAVING POLAR BEARS

Polar bears are in trouble. They live much of the year on sea ice, hunting their favourite food — seals. But our planet is getting warmer and their icy homes are melting away, forcing them to spend more time on land, scavenging for food in coastal towns and villages.

BEARS IN THE CITY

Other bear species have started moving closer to towns and cities, too. They've been spotted wandering onto school playgrounds, napping in trees, breaking into cars and even climbing through kitchen windows!

Brown bear raiding restaurant bins in Colorado Springs, USA

Amercian black bear grabbing a snack through a kitchen window in California, USA

American black bear taking a dip near holidaymakers in Lake Tahoe, USA

Spotting Polar Bears

WHERE TO SEE THEM

Churchill is one of the best places in the world to see wild polar bears. Specially designed bear-proof buggies take tourists on a bear hunt in the icy wilderness beyond the city limits. The buggies have a steel mesh floor, so you can see the bears up close as they wander underneath.

WHEN TO SEE THEM

The best time is in October and November as they travel to their winter hunting grounds.

LISTEN OUT FOR ...

Hissing, snorting, growling, roaring and chuffing (which sounds a bit like a steam engine).

The best time to see the Austin bats is in August, when the baby bats are born and the population doubles.

Bats use their wings to fly and also to climb and crawl.

Bats are strong flyers and can cover more than 50 km (30 miles) per night in search of food.

Wide ears that point forward

Furry body

Long tail and long, narrow wings for speedy flying and aerial manoeuvres

If you stand too close, you might be bombarded with bat droppings. Visitors are advised to wear a jacket and a hat!

LATIN NAME *Tadarida brasiliensis* (also known as the Brazilian free-tailed bat)
FAMILY Molossidae (free-tailed bats)
WINGSPAN 30–35 cm (11.8–13.7 in)
CONSERVATION STATUS
Least Concern
WHERE IN THE WORLD?
Parts of North and South America

Mexican Free-Tailed Bats

in Austin, USA

AS THE SUMMER SUN SETS OVER THE CITY OF AUSTIN, crowds of people gather on the banks of the lake overlooking Congress Avenue Bridge. They're here to watch as thousands of bats emerge from underneath the bridge, like great chattering plumes of smoke, to hunt insects in the night skies. On some evenings there may be as many as 1.5 million bats, flying at speeds of more than 100 kph (62 mph)!

Bats are the only mammals that truly fly (by flapping their wings, rather than by gliding).

Mother bats can carry the pups when they fly.

If you come across a grounded bat, you should not try to pick it up.

You can take a boat tour or paddle a kayak to the bridge to watch the bats from the water.

City Life

In 1980, Austin's Congress Avenue Bridge was redesigned, creating lots of tiny nooks and crannies on the underside that were perfect for bats to snuggle into. They began to move in in their thousands, and at first, the local residents weren't happy. Some even tried to have the bats removed. But now the people of Austin love their bats! It helps that the bats eat enormous numbers of insects. Together they can polish off up to 9000 kg (20,000 lb) of bugs each night, including mosquitoes and insect pests that destroy crops.

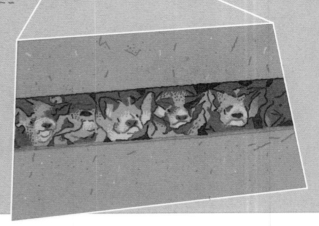

By day, the bats snooze in crevices on the underside of the bridge. At night, they scour the city for insects.

BAT HOMES

A bat's home is called a roost. Caves are the best-known bat roost, but many species happily make their homes in church roofs, bridges, barns, garages, sheds, cellars and other buildings.

In Sydney, Australia, thousands of grey-headed flying foxes (a type of bat) roost in large noisy tree colonies in the Royal Botanic Garden. At sunset, they fly across the city to feed on figs, nectar and pollen.

Each winter, around 10,000 bats cosy up inside Spandau Citadel, an old fortress in Berlin, Germany. Bats have been roosting here since the 16th century and once lived alongside medieval knights and kings!

One family of lesser horseshoe bats have set up home inside the belly of a *Triceratops* in a dinosaur theme park in Devon, UK!

BATS IN THE LIBRARY

In Biblioteca Joanina — a grand 18th-century library in Coimbra, Portugal — common pipistrelle bats are helping to stop ancient manuscripts and books from being devoured by book-eating insects. In the daytime, the bats sleep behind the bookshelves. At night, they swoop through the rooms to feast on bugs.

The bats have been living in the library since at least the 1800s.

BAT BOMBS

During the Second World War the US Air Force hatched a plan — known as Project X-Ray — to use bats to carry small incendiary (fire) bombs that would be dropped into Japanese cities. Luckily, the idea was scrapped before it was carried out.

FLYING IN THE DARK

Finding your way around in the dark isn't easy, but bats have an amazing trick called echolocation. They let out very high-pitched noises (too high for humans to hear) then listen to the echo as it returns from nearby objects. This helps them to work out where things are and to dodge obstacles.

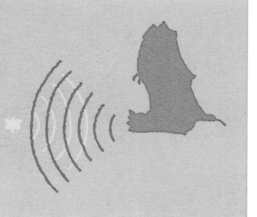

SPEEDY BATS

Mexican free-tailed bats are the jets of the animal world. They have been clocked at speeds of up to 160 kph (99 mph) — faster than any living bird in horizontal flight!

THE TRUTH ABOUT BATS!

Bats may have a reputation as blood-sucking monsters, but they are gentle, clever creatures and can be very useful. They are good at getting rid of insect pests, they pollinate flowers, and their droppings make a great fertilizer.

Spotting Bats

TOP TIP
Look out for bat droppings (known as guano). Guano is made up of dried insects and is usually quite crumbly.

SAVING BATS

The world can be a dangerous place for these furry little mammals. In some countries whole colonies of bats are killed in the mistaken belief that they are all vampires. Others are hunted for their meat or for medicine. It's time to stand up for bats!

WHERE TO SEE THEM

Close to their roosts or near their insect hunting grounds — flying over treetops, near rivers and lakes, or in your own garden. Some people are even lucky enough to have them roosting in the roof of their house.

WHEN TO SEE THEM

Bats hibernate in the winter and are most active in the summer when they give birth and raise their young. The best time is around sunset when the bats wake up and leave their roost to hunt.

LISTEN OUT FOR ...

'Click-click-click' — a bat's high-pitched chatter. If you have a bat in your attic you might hear scratching or flapping sounds as it moves about.

RACCOONS
in San José, Costa Rica

The city of San José is surrounded by mountains and volcanoes.

The super-talented raccoon can swim, dig, jump and climb pretty much anything.

Raccoons' front paws are so nimble they can pry the lids off rubbish bins, open windows and fridge doors and unscrew jars.

Raccoons have excellent night vision, making it easy for them to go on night-time raids across the city.

They can squish down their spines, so they can squeeze into small spaces, such as under garage doors.

Black burglar-style 'mask' over its eyes

Long, slender fingers and lightning-quick paws for grabbing small prey or snatching hamburger buns

SOME SEE THEM AS SCRAPPY HEROES, OTHERS AS TRASH-STEALING BANDITS, but there's no denying that the raccoon is one of the most successful urban-dwelling animals on the planet. As the lights go down in Costa Rica's packed capital, they hang around the neighbourhoods in gangs, raiding homes and restaurants, and sneaking into attics. They've even been spotted racing up the side of skyscrapers!

LATIN NAME *Procyon lotor*
FAMILY Procyonidae (mammal family that includes raccoons, coatis, cacomistles, kinkajous, olingos, olinguitos and ringtails)
LENGTH 40—60 cm (1.3—2 ft) plus 25—35 cm (0.8—1 ft) tail
CONSERVATION STATUS Least Concern
WHERE IN THE WORLD? North America. They have also been introduced to parts of Asia and Europe and now thrive there too.

City Life

Raccoons have adapted so well to life in the city, they are now more common here than in the countryside. They can be found in almost every major town in North America from Panama City to New York and Toronto — the raccoon capital of the world! They break into houses through cat flaps, ride the subway and have shown up in some funny places all over town…

Camping out above the baggage claim area at Toronto Pearson International Airport, Canada

Watching a baseball game in Baltimore, USA

Scaling to the top of a 25-storey office tower in St Paul, USA

ADAPTING TO THE CITY

Life in the city is a breeze for resourceful raccoons. They're known as 'trash pandas' because they'll eat pretty much anything, from grubs, fruit and berries to scraps left in rubbish bins. They are also happy to cosy up in attics, chimneys, garages, sheds, abandoned cars, and crawl spaces under buildings.

CITY SMARTS

Raccoons are super-smart, ranking higher than cats and just below monkeys on the IQ scale. Some scientists think that city life is making them even smarter. Urban raccoons have learned to topple over bins to break open the lids, unlatch locks and use sewer systems as highways to get around town.

They can even turn on the tap for a drink of water!

LIVING WITH RACCOONS

They may look cute, but raccoons are wild animals and sometimes carry diseases. Here are some tips on how to get along with your furry neighbours.

- Never offer them food and don't leave pet food outside — raccoons love cat food!
- Put bungee cords on your bins to secure the lids in place.
- Never try to pick up a raccoon or invite one into your home.
- If you see a raccoon that might need help, call your local wildlife centre and ask them for advice.

Spotting Raccoons

WHERE TO SEE THEM

Pretty much anywhere they might find food or shelter.

WHEN TO SEE THEM

They usually come out at night but can also be spotted in the day.

LISTEN OUT FOR…

Raccoons can make over 50 different sounds, ranging from soft purrs, to whistles, growls and hisses.

BROWN-THROATED SLOTHS
in Panama City, Panama

Panama's Metropolitan Natural Park has more than 300 species of animals including monkeys, anteaters, toucans and iguanas.

Three long claws on each foot — perfect for hooking onto branches and vines

Treasure-filled Spanish galleons used to sail into the city's ports.

Very small head for the size of its body

Green algae grows on its fur helping it blend in with the trees.

Sloths have upturned mouths, so they look like they're always smiling!

Sloths are one of the slowest-moving animals on the planet. They can spend around 90% of their time completely motionless!

PANAMA CITY WAS ONCE A HOTBED OF PIRATES FOLLOWING THE SPANISH TREASURE TRAIL ACROSS THE AMERICAS. Today, it is a super-modern metropolis of shimmering glass skyscrapers, steel towers and traffic jams. It is also one of the few cities in the world where you might see a sloth smiling down from the treetops, hanging on washing lines or crawling across the street!

LATIN NAME *Bradypus variegatus*
FAMILY Bradypodidae (three-toed sloths)
LENGTH 42–80 cm (1.4–2.6 ft)
CONSERVATION STATUS
Least Concern
WHERE IN THE WORLD?
Parts of Central and South America

City Life

Jungle-loving sloths are one of the last animals you might expect to see in a busy town, but right in the middle of Panama City is an island of lush tropical rainforest — the Metropolitan Natural Park. The sloths spend most of their time munching leaves in the park, but sometimes they lose their way and end up in the city. Some even move into people's back gardens.

SLOTHS IN TROUBLE

The city contains many dangers for sloths. Shocks from power lines can be deadly and navigating traffic can be treacherous. It can take them more than three minutes to cross the street!

When new roads slice through the sloths' forest homes, the sloths have to crawl across the streets on their tummies to get to the trees on the other side.

GETTING AROUND

Their long claws and weak back legs can make walking on the ground very difficult. But sloths have some special tricks for getting around the city.

They are excellent climbers and can clamber along garden fences.

Their curved claws are perfect for hanging on washing lines.

They use power lines like bridges to cross busy roads.

They usually leave the treetops just once a week to go to the toilet!

LIFE IN SLOW MOTION

The secret to the sloth's slowness lies in its leafy diet. Leaves are very low in energy. Sloths cope with this by eating A LOT of leaves and by doing as little as possible. When they do get going, they move VERY slowly.

Spotting Sloths

SLOTH RESCUES

Wildlife rescuers are often called out to help return stranded sloths to their tree homes. If you see a sloth, don't try to pick it up. They can bite and those claws are sharp! Best to leave it to the experts.

Sloths have a vice-like grip and can get feisty when threatened!

WHERE TO SEE THEM

In the highest branches of rainforest trees. If they're lost in the city, they could be hanging from power lines or crossing roads.

WHEN TO SEE THEM

You can spot them in the trees any time of day and all year round. Stay nice and quiet and go with a guide who knows the area.

LISTEN OUT FOR ...

'Ahh-eeee' — the sloth's long, high-pitched call that echoes through the forest

'Squeak' — the cry of a baby sloth calling out for its mother

RED SQUIRRELS
in St Petersburg, Russia

IT'S WINTER IN ST PETERSBURG. A BLANKET OF SNOW COVERS THE CITY and colourful churches with onion domes dazzle like Christmas tree ornaments. Many mammals snooze through the colder winter months — but not the hardy red squirrel. Head to one of the parks or gardens with a handful of nuts and they'll scamper over your feet and eat straight from your hands!

Many of the seeds that squirrels forget to retrieve grow into trees.

LATIN NAME *Sciurus vulgaris* (Eurasian red squirrel)

FAMILY Sciuridae (family of small rodents that includes tree squirrels, ground squirrels, chipmunks, marmots, flying squirrels and prairie dogs)

LENGTH 19—23 cm (7.5—9 in) plus 15—20 cm (6—8 in) tail

CONSERVATION STATUS Least Concern

WHERE IN THE WORLD? Europe and Northern Asia

Squirrels can sniff out food buried under the snow.

DID YOU KNOW?

Squirrels don't only munch on nuts and seeds — they eat berries, mushrooms, flowers, bugs and even bird eggs and baby birds.

Long bushy tail for balancing and climbing

Squirrels are expert climbers. They can run headfirst down trees and hang upside down by their back feet.

Sharp claws for scurrying up and down trees

Pointed ears that grow fluffy tufts in the winter

Cheek pouches for storing food

Squirrels have four front teeth that never stop growing, so they can nibble away as much as they like!

43

City Life

St Petersburg is the greenest of Russia's big cities. It has hundreds of parks and gardens, tree-lined streets and leafy squares. So, there are lots of places for squirrels to build their homes and a plentiful supply of nuts, seeds and other squirrel snacks.

St Petersburg's parks are a popular squirrel hangout.

SQUIRREL HOMES

Red squirrels build ball-shaped nests made of twigs and leaves and lined with moss. You can spot them in tree forks or hollows. Some don't bother making nests of their own and move into woodpecker holes or nest boxes meant for birds.

A squirrel's nest is called a drey.

Road signs remind drivers to 'Mind the Squirrels!'

ОСТОРОЖНО, БЕЛКИ!

CITY HAZARDS

There are downsides to living in cities. Crossing busy roads can be a dangerous business for squirrels. And they have to beware of hungry household pets, especially cats.

TAKING CARE OF SQUIRRELS

Specially designed squirrel feeders and nest boxes can help stop sneaky squirrels from pinching food and shelter from birds. Hang them high in a tree or on a fence so the squirrels can hide from predators.

Make sure the feeder or nest box is in a place where there are no cats nearby.

SECRET LANGUAGE OF SQUIRRELS

You probably know that red squirrels use their bushy tails to keep their balance. But did you know they also use them to talk to one another? If you see a pair of squirrels with fluffed-up tails, stomping their feet, they're probably having an argument!

A wag of the tail means the squirrel is startled or alarmed.

NUTS ABOUT NUTS

Squirrels don't hibernate through the winter. They collect nuts, seeds and berries in the autumn then 'squirrel' them away in tree crevices or underground larders to snack on later.

Squirrels' cheeks are like massive grocery bags. When they have a full load, they take it back to their storehouses.

MULTIPURPOSE TAILS

Next time you see a squirrel, take a look at what it's doing with its tail. They use their amazing tails for all kinds of things ...

As an umbrella (to shelter from the rain or keep cool in the sun)

As a parachute (handy when making a quick exit from a tree)

As a fluffy duvet (to keep it warm in the winter)

SQUIRRELS OF THE WORLD

Other squirrels and their cousins have set up home in cities, too ...

In North America, grey squirrels sometimes move into attics, chimneys or basements to escape the winter cold and raise their young.

In Ontario, Canada, ground squirrels known as woodchucks build underground burrows in backyards and on golf courses.

According to scientists in Ontario, Canada, eastern chipmunks living in cities (where they have easy access to food) are fatter and happier than those that live in the countryside.

Spotting Squirrels

TOP TIP
If you see a squirrel rushing about, it's probably collecting or hiding its nibbles!

TOP TIP
Look out for gnawed pine cones and piles of squirrel droppings near feeding sites.

WHERE TO SEE THEM

Squirrels are a common sight in parks, squares and gardens in towns and cities around the world. Spot them scampering about, eating nuts and dashing up trees.

WHEN TO SEE THEM

Early morning and late afternoon as they search for food.

LISTEN OUT FOR ...

Squeaks, bark-like grunts or scratching noises that can be heard as they dart about in roof spaces.

REPTILES and AMPHIBIANS

REPTILES AND AMPI

THERE ARE THOUSANDS OF SPECIES OF REPTILES AND AMPHIBIANS LIVING ON OUR PLANET and a surprising number can be found right in the middle of busy cities. They are very good at blending in with their surroundings, but open your eyes and ears and you may discover lizards dancing on rooftops, frogs croaking beside garden ponds and armies of toads crawling across busy highways!

Reptile or Amphibian?

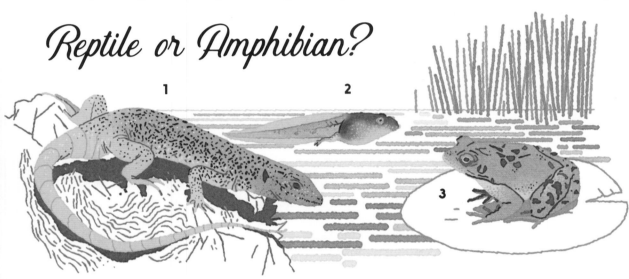

1. Reptiles breathe air and have dry, scaly skin. The main types of reptile are snakes, crocodiles and alligators, turtles, tortoises and lizards.

2. Most amphibians lead double lives. When they're young they live in the water. They breathe through gills and have fins to help them swim. The young are often known as tadpoles.

3. When they're older, amphibians grow lungs, and can breathe air and live on land. Amphibians include frogs, toads, salamanders, newts and caecilians.

Tracks

Spotting tracks left by reptiles and amphibians can be tricky. Some are quite small and their tiny footprints are hard to see. Others leave behind marks made by their long tails or bodies!

1. **Frogs and toads**
   Frogs and toads have four toes on their front feet and five toes on their back feet. Frogs usually jump, while toads usually walk.

2. **Lizards**
   Lizards leave behind scratch marks from their feet and claws, and drag marks from the swoosh of their tails.

3. **Crocodiles and alligators**
   Crocodile and alligator tracks look a bit like a lizard's but are usually bigger!

4. **Snakes**
   Snakes don't have feet — they leave marks where their body has wriggled across the ground.

5. **Turtles and tortoises**
   Sea turtles drag themselves along with their flippers. Their prints look like mini bulldozer tracks.

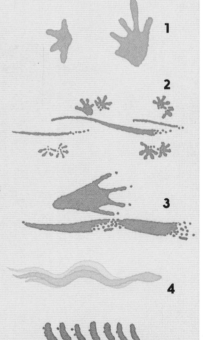

New clothes

As reptiles and amphibians get bigger, their skin or scales become too small for them. So, they get rid of their old clothes and grow new ones.

1. **Frogs and toads**
   Some frogs and toads change their skin once a week. Others do it every day. Then, if they're peckish, they will eat the old dead skin!

2. **Lizards**
   Some lizards remove their old skin all in one go. Others shed it bit by bit at different times.

3. **Crocodiles and alligators**
   Crocodiles and alligators are a bit different. They shed individual scales — so you won't find an old crocodile-suit lying around!

4. **Snakes**
   Snakes crawl out of their old skin, leaving it behind in one piece.

5. **Turtles and tortoises**
   Turtles and tortoises shed both their skin and the scales that cover their hard shell.

BIANS IN THE CITY

Eggs

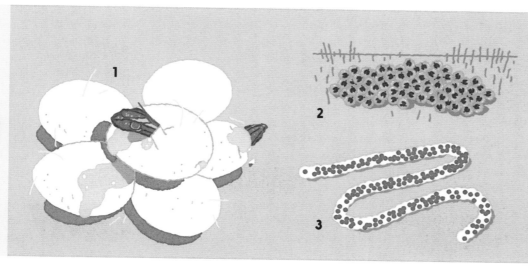

Eggs are another clue left behind by reptiles and amphibians.

1. Snake eggs

Although some reptiles give birth to live young (like mammals do), most lay eggs. Many are soft and leathery, but some are hard like bird eggs. Look out for old egg casings hidden under leaves.

2. Frogspawn

Frogs lay their eggs in big jelly-like clumps.

3. Toadspawn

Toads lay long strings of eggs — a bit like a necklace.

WATCH THEM GROW

Check ponds and puddles for frogspawn and toadspawn (eggs) floating on the surface or clinging to weeds. If you visit regularly you can watch them transform from tiny tadpoles into grown-up frogs and toads!

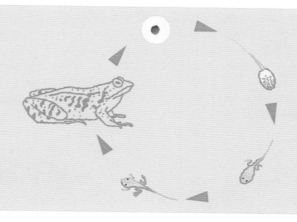

HIDING PLACES

Here are some popular reptile and amphibian hangout spots:

- In the undergrowth in gardens, parks and roadside verges.
- Near piles of wood or rocks, pavement cracks or wall crevices.
- In sewers, ponds, lakes, streams and puddles.

RAINY DAYS

Turtles, frogs and toads love a good rainstorm. It's a great time to snack on worms and splash about in warm puddles. Next time it rains, get your wellies on and join them!

SUNBATHERS

Reptiles and amphibians are cold-blooded animals. This means they can't make their own heat, so they warm themselves by basking in the sun. If they get too hot, they take a quick dip in the water or cool off in the shade — like we do on a sunny day! See if you can spot them sunbathing on walls, fence posts or pavements.

FROG CHORUS

Keep your ears open. Male frogs are some of the chattiest animals around — especially when they're looking for love. Each species has its own unique call — from high-pitched whirrs, chirps and clicks, to ribbits, hoots, bonks and barks. You can identify the type of frog just by listening.

Some frogs have throat pouches that stretch like balloons when they sing.

Jewelled Chameleons
in Antananarivo, Madagascar

THE ISLAND OF MADAGASCAR IS ONE OF THE MOST BIODIVERSE PLACES ON THE PLANET. It is home to all kinds of weird and wonderful animals from oddly shaped insects to cuddly ring-tailed lemurs. But its towns are growing, precious habitat is being lost and many of its unique species are in trouble. One hardy reptile battling it out in the capital city is the jewelled chameleon — famed for its bulging eyes, long sticky tongue and incredible colour-changing abilities!

Eyes that can rotate independently of each other, allowing it to search for prey withou moving its body

Large toes that help it cling on tight to branches

Skin that changes colour from light green to brown, with small bright spots

Chameleons are insect eaters and are very good at getting rid of insect pests.

City Life

LATIN NAME
Furcifer lateralis
FAMILY Chamaeleonidae (chameleons)
LENGTH 17–25 cm (7–9 in)
CONSERVATION STATUS
Least Concern
WHERE IN THE WORLD?
Madagascar

Animals such as chameleons can make do with small patches of forest in cities where larger animals, such as lemurs, would long since have disappeared. In Antananarivo, the jewelled chameleon can be spotted in the Tsimbazaza zoological park in the middle of the city, in gardens with lots of plants and in clumps of vegetation next to roads.

Instead of wrestling each other, male chameleons have fierce colour-changing competitions.

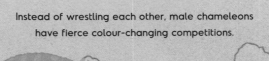

The winner is the one with the most impressive colours!

CHANGING COLOURS

You might think that chameleons change the colour of their skin to match their surroundings. In fact, they're not hiding, they're communicating — letting others know what mood they're in, defending their territory, or showing that they're looking for love. An angry chameleon may turn from pale green to bright yellow, and to impress a mate, a chameleon might change into all the colours of the rainbow.

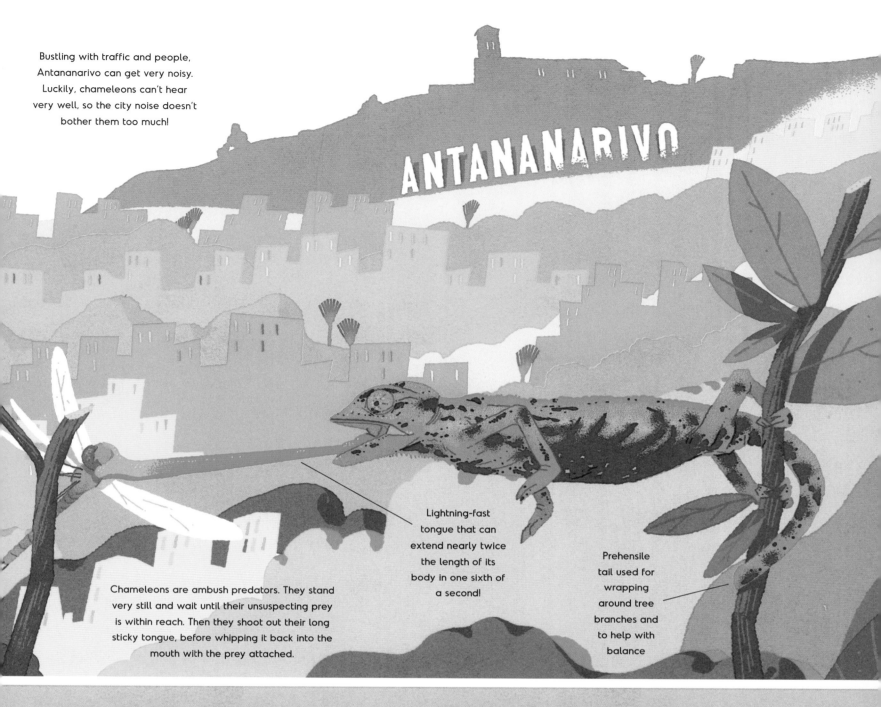

Bustling with traffic and people, Antananarivo can get very noisy. Luckily, chameleons can't hear very well, so the city noise doesn't bother them too much!

ANTANANARIVO

Lightning-fast tongue that can extend nearly twice the length of its body in one sixth of a second!

Prehensile tail used for wrapping around tree branches and to help with balance

Chameleons are ambush predators. They stand very still and wait until their unsuspecting prey is within reach. Then they shoot out their long sticky tongue, before whipping it back into the mouth with the prey attached.

Signs warn drivers to slow down so chameleons can cross roads safely.

CROSSING ROADS

Many people in Madagascar believe that killing a chameleon is *fady*, or taboo, and will bring bad luck. Drivers will do all they can to avoid running them over.

CHAMELEONS IN PERIL

Losing their forest homes isn't the only danger for chameleons. Each year, thousands are captured from the wild and sold as pets. Many then die as people don't know how to look after them properly.

In Morocco, some people think that chameleons have magical powers. They are sold in the city's souks (markets) and used to make potions or to perform magic spells.

Spotting Chameleons

WHERE TO SEE THEM

Chameleons are quite shy and prefer to hide out in trees, bushes and shrubs.

WHEN TO SEE THEM

During the daytime, when they're most active.

LISTEN OUT FOR...

Chameleons are mostly silent, but sometimes make hissing sounds.

WATER MONITORS
in Bangkok, Thailand

Water monitors usually live in burrows dug into the riverbanks. In Bangkok, concrete water pipes make ready-made homes.

Forked tongue for smelling its prey

Big claws and powerful legs that help it climb trees

Strong tail that it can lash out to protect itself and use to steer through the water

Scaly skin

LATIN NAME
Varanus salvator
FAMILY Varanidae (a lizard family that includes monitors and the Komodo dragon)
LENGTH 1.5–2 m (4.9–6.6 ft) or more
CONSERVATION STATUS
Least Concern
WHERE IN THE WORLD?
Southeast Asia

City Life

The ponds, lakes and trees of Lumpini Park make an ideal home for Bangkok's water monitors. They spend their days sunbathing on the grass alongside picnickers and dipping in and out of the water to hunt for their dinner. They don't seem to be put off by people or busy places; there are few natural predators, and there's lots of food for them to eat.

OUT AND ABOUT

Water monitors have also been spotted out and about in the city.

Slithering through the drains and sewers

Crossing busy roads in the middle of town

In the town of Chachoengsao, Thailand, a water monitor was found hiding under a cupboard in a local grocery store, eating a box of dried noodles!

FAMED FOR ITS GOLD-SPIRED TEMPLES, TASTY FOOD AND ZOOMING TUK-TUKS, Thailand's capital is one of the most popular tourist destinations in the world. Right in the middle of this megacity is Lumpini Park — an oasis of green where you can jog, cycle, paddle in the lake or relax on the grass. The park is also a popular hangout for around 400 water monitors — large reptiles that look like miniature dragons and can grow to more than 2 m (6.6 ft) long!

Water monitors look a bit like small alligators, but they are actually gigantic lizards.

Powerful jaws and sharp teeth

DINNER TIME

Water monitors are extreme carnivores and eat everything from fish, frogs and turtles to rodents, snakes, birds and monkeys. However, they don't usually attack humans. So long as you don't provoke them, you should be safe!

They help to keep the city clean by eating edible rubbish and rodents.

LUCKY LIZARDS?

Some people believe that water monitors are a bad omen, maybe because they hang around cemeteries and so are thought to bring bad spirits with them. Others think that if a water monitor enters your home it's good luck and if you find a water monitor inside your house, you should speak to it nicely to bring even more good fortune!

Spotting Water Monitors

WHERE TO SEE THEM

On or near riverbanks, ponds, canals or swamps. Hiding or basking in tree branches.

WHEN TO SEE THEM

They are active during the day and sleep at night.

LISTEN OUT FOR ...

A hissing sound. This means the water monitor feels threatened, so it's time to back away.

RETICULATED PYTHONS
in Jakarta, Indonesia

INDONESIA'S CAPITAL IS A MASSIVE SPRAWL OF AROUND TEN MILLION PEOPLE. Its glitzy skyline is crammed with skyscrapers and glamorous malls. On the edges of the city, meanwhile, densely packed slums overflow with rubbish. It's here, slinking through the streets and sewers, that you'll find Jakarta's most elusive inhabitants...giant pythons that can weigh more than the average adult human!

Jakarta has some of the tallest buildings in the world, including the 310 m (1017 ft)-tall Gama Tower.

LATIN NAME *Python reticulatus*
FAMILY Pythonidae (pythons)
LENGTH 3–9 m (10–30 ft) or longer
CONSERVATION STATUS Least Concern
WHERE IN THE WORLD? South and Southeast Asia

Jakarta sits on swampy land, criss-crossed with rivers. In the rainy season, snakes swim through the flooded streets to reach higher ground.

The reticulated python is the longest species of snake in the world and can grow as thick as a telegraph pole.

Pythons can't see very well, but they have special organs that sense the warm bodies of their prey. This helps them to hunt at night and in dark places.

They use their forked tongues to 'smell' the air and can feel vibrations through their bodies.

City Life

Reticulated pythons usually live in tropical forests and wetlands, but in Indonesia their forest homes are being destroyed, forcing them into nearby cities. In some parts of Jakarta, people leave rubbish outside their homes. This attracts rats and mice — the python's favourite foods.

The snakes dine out at the local rubbish dumps. Without snakes, the city would be overrun by rats.

KILLING MACHINES

A python doesn't inject venom into its prey like some snakes do. It grabs onto its victim with its sharp teeth, coils its body around it and squeezes it to death. Then it swallows it whole. In cities, pythons feast mainly on rodents, cats, dogs and chickens. Although this is rare, they have been known to kill humans. They can kill a person in minutes and swallow them in an hour!

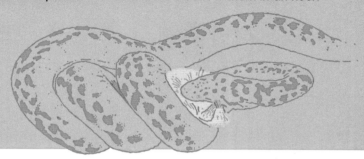

SNAKES OF THE WORLD

Jakarta isn't the only city in the world with snakes ...

In some tropical cities, snakes follow rats into sewers — and sometimes slither out of toilets!

Two male carpet pythons were discovered duelling in the attic of a house on the Sunshine Coast, Australia.

In Florida, USA, a Burmese python was seen wrestling an alligator on a golf course. No one's sure who won!

SNAKE RESCUERS

Some cities, including Jakarta, have expert snake catchers who you can call on to remove unwanted houseguests and return them to the wild.

HOW TO PREVENT A SNAKE FROM ENTERING YOUR HOUSE

1. Don't leave edible rubbish nearby. This will attract rats, and rats attract snakes.
2. Don't leave objects piled up outside your home. These are tempting hiding places for snakes.
3. Don't leave pet food outside — feed your pets inside your home.

Scaly waterproof skin to keep moisture in and stop the snake drying out in the heat

Spotting Snakes

WHERE TO SEE THEM

Hunting rats on rubbish dumps, lurking in sewers, in attics and backyards, schools, offices, shops and restaurants.

WHEN TO SEE THEM

At night, when they come out to hunt. Or during the rainy season, when heavy rains wash snakes out into the open.

LISTEN OUT FOR ...

Hissing sounds (the noise they make when they're disturbed), or sliding and banging noises from attics and wall spaces.

DON'T FORGET!

Not all snakes are dangerous, but some are deadly, so we don't recommend that you go looking for them!

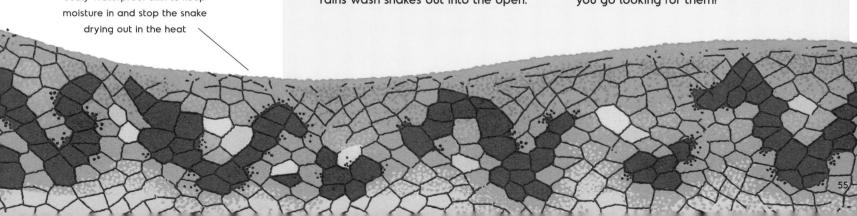

TOKAY GECKOS
in Manila, Philippines

WITH ITS JUMBLE OF SKYSCRAPERS, ENORMOUS SHOPPING CENTRES AND BUSY STREETS, Manila is one of the world's largest cities. This noisy metropolis may not seem like an obvious wildlife hotspot, but one colourful creature has made itself right at home here — the tokay gecko. This big-eyed lizard can be found all over the city clinging to walls inside houses and singing love songs into the night!

Large eyes for night-time hunting

Tiny hairs on its feet that allow it to cling to any surface

Blue/grey skin with yellow, orange or bright red spots

LATIN NAME *Gekko gecko*
FAMILY Gekkonidae (geckos)
LENGTH 30—35 cm (11.8—13.8 in)
CONSERVATION STATUS
Least Concern
WHERE IN THE WORLD?
Native to South and Southeast Asia; introduced to Madagascar and parts of North America

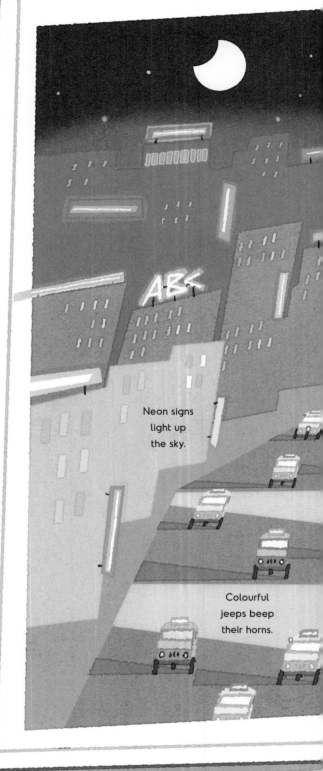

Neon signs light up the sky.

Colourful jeeps beep their horns.

DID YOU KNOW?

If attacked, the tokay gecko can detach the tip of its tail and grow a new one. The old part of the tail continues to twitch and thrash about. This distracts the attacker long enough for the gecko to make its escape!

City Life

In the wild, tokay geckos usually live in rainforests where there are lots of trees and other plants. But they have adjusted well to life in the city. Thanks to the millions of tiny hairs on the bottom of their feet, they can grip onto even the smoothest surfaces. So they have no trouble running up concrete walls, glass windows and lamp posts. They can even hang upside down from ceilings.

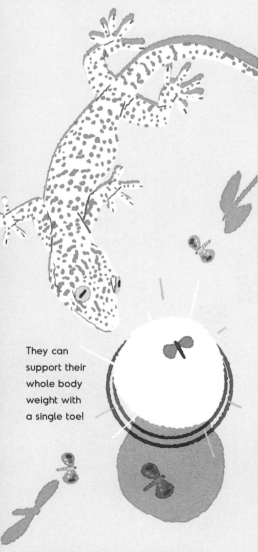

They can support their whole body weight with a single toe!

Manila has one of the brightest night skies in the world. And bright lights attract lots of insects such as moths — one of the gecko's favourite snacks.

BUG CATCHERS

Tokay geckos have big appetites and eat all kinds of household bugs including spiders, cockroaches and swarms of termites. They may also eat small rats and mice. So they're handy to have around!

TINY DRAGONS

In Asian folklore, tokay geckos are thought to be the direct descendants of dragons. They're said to bring great luck and prosperity.

They have a sticky tongue to capture their prey and to clean away dust and dirt from their eyes.

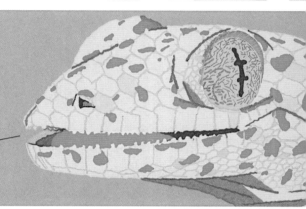

GECKO V SNAKE

Tokay geckos are very feisty and have been spotted battling with snakes (one of their main predators) on the streets of the city.

Spotting Geckos

SAVING GECKOS

Some people wrongly believe that tokay geckos can cure different illnesses, so many are captured and traded illegally for medicine. Because of this they are becoming rare and may start disappearing from many cities altogether.

WHERE TO SEE THEM

Climbing on walls, ceilings and windows inside houses or on the outsides of buildings. Look around window screens and close to porch lights where moths and insects gather. But don't get too close! The tokay gecko is one of the biggest geckos and has a very sharp bite.

WHEN TO SEE THEM

Geckos can be seen any time of the year. They usually hunt at night so that's the best time to see them.

LISTEN OUT FOR ...

'Tokay-tokay' — the loud noise the males make to attract a mate (this is where they get their name from). They also make barking, clicking and chirping sounds to talk to one another. When they're attacked they make a hissing or croaking noise.

LOOK OUT FOR ...

Shed skins: like other lizards, tokay geckos change their skin as they grow. First they break the skin at the head then strip off towards the tail.

ANDEAN MARSUPIAL FROGS
in Quito, Ecuador

Andean marsupial frogs don't lay their eggs in the water. The female keeps them safe in a little pouch on her back until they hatch into tadpoles.

Colourful houses hug the hillside.

From Quito, tourists can explore two of the most amazing wildlife destinations on the planet — the Amazon jungle and the Galápagos Islands.

Long sticky tongue for slurping up insects

Camouflaged skin for hiding among the leaves

Powerful legs for leaping and webbed back feet for swimming

SITTING BETWEEN SOARING PEAKS IN THE ANDES MOUNTAINS, QUITO IS ONE OF THE HIGHEST CITIES IN THE WORLD. Its hills and ravines were once a paradise for frogs and toads. They could be found hopping around parks, hunting insects near street lights and croaking out mating calls from ponds and puddles. But habitat loss and pollution have left many species fighting for survival...

LATIN NAME *Gastrotheca riobambae*
FAMILY Hemiphractidae (a family of frogs from Central and South America)
LENGTH 43–48 mm (1.7–1.9 in)
CONSERVATION STATUS Endangered
WHERE IN THE WORLD? Ecuador

City Life

The Andean marsupial frog used to be a common sight in Quito, but in recent years its numbers have declined, and it's now endangered. Luckily, help is at hand! Conservation groups have been working hard to create a more frog-friendly habitat in one of the ravines next to the Viva Ruta — a big road that runs around the city. So far, around 500 frogs have been reintroduced to the area.

The frogs love plants with big leaves to shelter under and flowers that attract insects. Ponds provide a place for tadpoles to grow up.

FROGS OF NEW YORK

Cities aren't all bad news for frogs. A brand new species, known as the Atlantic Coast leopard frog, has been found living in swamps of New York City, USA!

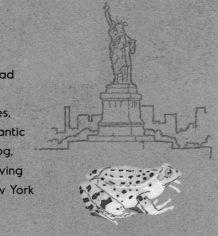

WHY DID THE FROG CROSS THE ROAD?

Frogs and toads move from place to place to find food or to get to breeding sites. These journeys can be perilous — especially if there are busy roads to cross. Sometimes, hundreds can be wiped out in one night as they attempt to hop and dodge through the speeding traffic.

We can help by building new roads away from frog migration routes, or temporarily closing roads so frogs can cross safely.

In Cape Town, South Africa, drivers are reminded to slow down and watch out for Western leopard toads crossing.

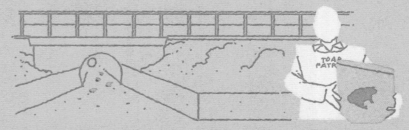

Tunnels under roads offer a safe route to the other side.

Volunteer patrollers give frogs and toads a helping hand by carrying them across the roads in buckets.

SINGING LOVE SONGS

In Taiwan, tiny mientien tree frogs have a nifty way of making themselves heard over the noisy city traffic. They gather in storm drains under the streets, then sing their hearts out! They seem to have worked out that the concrete tunnels make their mating calls louder.

The louder their call, the more likely they are to impress the ladies!

Spotting Frogs

TOP TIP
Look for tadpoles in shallow pools.

WHERE TO SEE THEM

Hiding in vegetation near pools, streams, lagoons, swamps, rivers, garden ponds, puddles and roadside ditches.

WHEN TO SEE THEM

They're most active at night, when they move from place to place, or after it rains.

LISTEN OUT FOR...

'Cluck-cluck-cluck' — the mating call of the male Andean marsupial frog

BUGS

BUGS IN THE CITY

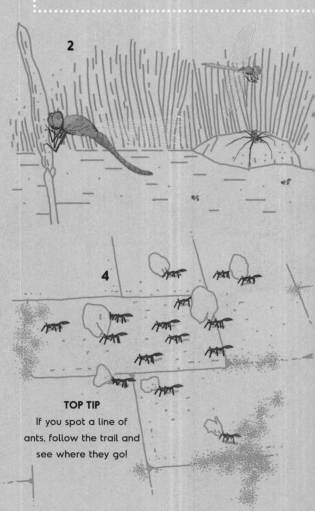

FROM TEENY ANTS TO BEAUTIFUL BUTTERFLIES, BILLIONS OF BUGS SHARE OUR CITIES. They're all around us — scurrying along pavements, digging in flower beds and zooming through the sky. Most are pretty small, but they are easy to find. You just need to stop and look around you, and be careful where you step!

ALL KINDS OF BUGS

These little creatures belong to a group called invertebrates, which includes insects, spiders, snails, earthworms, centipedes and microscopic mites.

TOP TIP
Count the legs. If it has six legs, it's an insect. Spiders have eight legs and earthworms have none.

URBAN BUG SAFARI

Bugs don't leave behind big footprints like birds and mammals do, but you don't need to go far to discover their amazing miniature worlds. Head to the nearest tree or tuft of grass, or turn over a few rocks, and you're bound to find something.

REMEMBER
Compared to these tiny creatures, you're a giant. Be gentle and try not to scare them.

1. In among the flowers
Your own back garden or local park is a great place to start your bug safari. Crouch down low, stay nice and still, then see how many creeping, crawling creatures you can spot.

2. By the waterside
Lakes, rivers and ponds are a popular hangout for dragonflies, fishing spiders, mayflies, pond skaters and water beetles. See them skimming above the water or dip a net under the surface and see what you can find.

3. Walls and fences
Slugs and snails are renowned for their incredible climbing skills. You can find them travelling up walls and fences or clinging to garden furniture. The best time to find them on the move is at night or after it's rained.

4. Nooks and crannies
Bugs can get into very small spaces and many of them love to lurk in dark, damp places. Check under logs and rocks, beneath plant pots, and in cracks in walls and pavements.

TOP TIP
Don't forget to check under leaves for caterpillars.

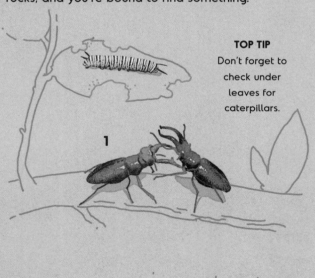

TOP TIP
If you spot a line of ants, follow the trail and see where they go!

Moths, crane flies and some beetles are attracted to artifical city lights.

BRIGHT CITY LIGHTS

For night-flying insects that usually navigate by moonlight, artificial lights can be confusing. That's why you often see them buzzing around the glow of street lamps. Go outside with a torch after sunset, then watch to see what turns up.

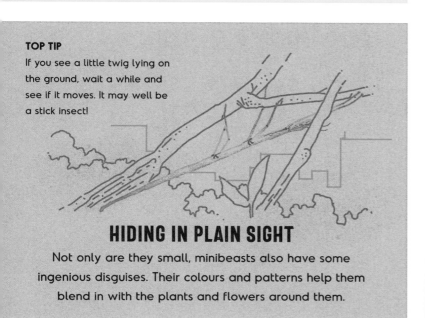

TOP TIP
If you see a little twig lying on the ground, wait a while and see if it moves. It may well be a stick insect!

HIDING IN PLAIN SIGHT

Not only are they small, minibeasts also have some ingenious disguises. Their colours and patterns help them blend in with the plants and flowers around them.

BUG HIGHWAYS

Don't forget to look up. There are millions of flying critters zooming over our cities. Next time you see one passing you by, try to work out what kind of creature it is and think about where it might be going.

MINI HOUSEGUESTS

Some minibeasts like to hibernate in buildings for the winter, some live there all year round. Others just pop inside to look for food. The most common ones you'll find in your home are flies, ants, beetles, bees and spiders, but in some cities, house guests include scorpions, cockroaches and tarantulas!

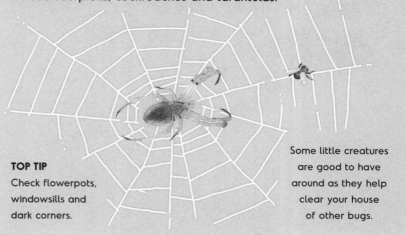

TOP TIP
Check flowerpots, windowsills and dark corners.

Some little creatures are good to have around as they help clear your house of other bugs.

INSECT ORCHESTRA
Listen out for crickets singing late at night.

GOOD CITIZENS

They may be tiny, but insects and other minibeasts provide an important public service by munching their way through decaying matter and by pollinating plants. Without them our cities would be very messy and many of our favourite flowers, fruits and vegetables wouldn't be able to grow. They are also tasty snacks for many different animals.

DRAGONFLIES
in Lisbon, Portugal

Transparent wings that reflect light to create the colours of the rainbow

Six legs that it uses to scoop up prey in mid-air

Long, thin body

Massive compound eyes made up of thousands of eyelets (tiny eyes)

Dragonflies can fly at 100 body-lengths per second. That's the same as a person travelling 150 m (328 ft) in the blink of an eye.

A dragonfly has huge eyes — so big they take up nearly all of its head. It can see in almost every direction and spot insects from far away.

Very strong jaws for crushing its prey

There are many different species of dragonflies in Lisbon including the red-veined darter and the emperor dragonfly (shown here).
WINGSPAN 17–191 mm (0.6–7.5 in)
CONSERVATION STATUS
Varies depending on species
WHERE IN THE WORLD?
Every continent except Antarctica and Greenland

City Life

Dragonflies love water. They spend most of their lives underwater as larvae munching on tiny fish, tadpoles and mosquito larvae. As adults, they lay their eggs on plants in or close to water and hunt nearby. City waterways — from ditches and ponds to large lakes and rivers — can make the ideal dragonfly home, so long as the water is clean and there are lots of insects for them to snack on.

Lakes built on rooftops create dragonfly-friendly habitats in the middle of cities.

THE IMPORTANCE OF DRAGONFLIES
There are lots of reasons to be happy to see a dragonfly!

1. They need pure, clean water to survive. Their disappearance from a pond could be a sign that the water has been polluted.
2. They are ferocious predators and eat insects that bite us, such as midges and mosquitoes. Sometimes they are released into places infested with mosquitoes to help stop the spread of diseases.
3. They make a tasty meal for other animals such as frogs and birds.

WITH ITS PRETTY COBBLED STREETS AND SANDY BEACHES, LISBON ATTRACTS VISITORS FROM ALL OVER THE WORLD. It is also a popular stop-off for globetrotting dragonflies, heading south to Africa for some winter sun. Keep your eyes peeled towards the end of summer and you might spot thousands of these shiny big-eyed flying machines streaming across the city to hunt insects in the rivers, lakes and ponds.

In Japan, it was believed that the spirits of dead ancestors would return on the backs of dragonflies. In Ireland, they were thought to be horses for fairies!

Dragonflies are usually found close to water but have been spotted flying past office buildings in the middle of the city. Sometimes they get confused by clear glass and crash straight into windows.

Dragonflies are champion flyers and can migrate long distances across the sea. They can also hover like a helicopter and zoom from left to right, up, down, forwards and backwards.

WATCHING DRAGONFLIES

If you are really lucky, you may get to see a young dragonfly transform from underwater nymph (larva) to dazzling flying machine. When the larva is ready to change into an adult, it crawls out of the water, cracks open its hard skin and wriggles out. It takes breaths of air to plump out its body, unfolds its wings and waits for them to dry. Finally, it's ready for its first flight!

Spotting Dragonflies

WHERE TO SEE THEM

Resting on reeds next to rivers, streams, lakes or ponds, or darting just above the surface.

WHEN TO SEE THEM

On warm, sunny days.

LISTEN OUT FOR ...

A deep humming noise as it beats its wings.

TOP TIP

Dragonflies warm themselves up by basking in the sun. If you spot one perching, approach it slowly and quietly so you don't scare it away.

TREE WĒTĀ
in Wellington, New Zealand

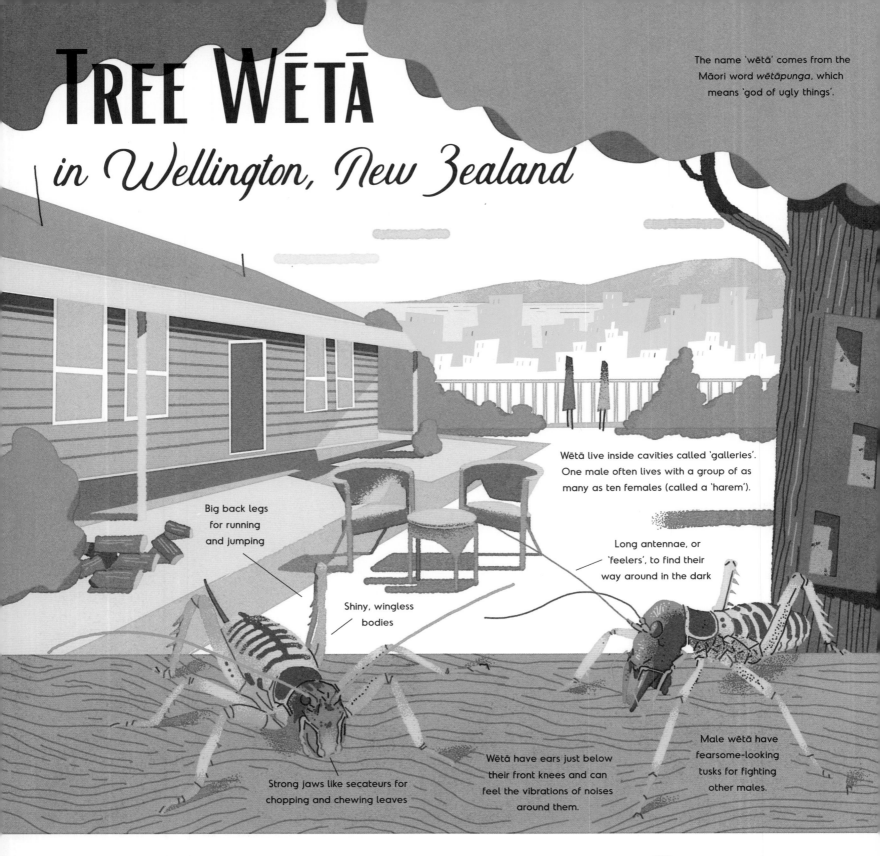

The name 'wētā' comes from the Māori word *wētāpunga*, which means 'god of ugly things'.

Wētā live inside cavities called 'galleries'. One male often lives with a group of as many as ten females (called a 'harem').

Big back legs for running and jumping

Long antennae, or 'feelers', to find their way around in the dark

Shiny, wingless bodies

Strong jaws like secateurs for chopping and chewing leaves

Wētā have ears just below their front knees and can feel the vibrations of noises around them.

Male wētā have fearsome-looking tusks for fighting other males.

NEW ZEALAND'S CAPITAL IS ONE OF THE BEST CITIES IN THE WORLD FOR NATURE SPOTTING. Pods of dolphins and orcas hang out in the harbour, little blue penguins shuffle across the streets to reach their nests, and gangs of gulls lurk on the quayside looking for sandwiches. Meanwhile, giant insects the size of mice hide out in suburban shrubberies. With their curved tusks and spiny legs, tree wētā may look terrifying, but they're probably more scared of you than you are of them!

LATIN NAME *Hemideina crassidens* (Wellington tree wētā)
FAMILY Anostostomatidae (a family of insects that also includes king crickets and giant wētā)
BODY LENGTH 4–7 cm (1.5–2.8 in)
CONSERVATION STATUS Least Concern
WHERE IN THE WORLD? New Zealand

City Life

Wētā are endemic to New Zealand, which means they only live here. There are more than 70 different species, but the most streetwise are the tree wētā, which can be found in back gardens in almost every town in New Zealand. They spend their days hiding out in holes inside trees or fallen logs, in building crevices or garden sheds, in the folds of corrugated iron, or even inside welly boots!

New Zealanders know to always shake out their wellies before putting them on!

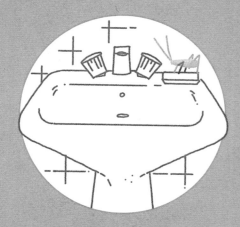

Sometimes wētā wander inside people's houses. If you find one in your home, release it into the garden next to a leafy bush.

When a wētā is feeling scared, it raises its long spiky legs in the air to make itself look as big as possible.

WĒTĀ V CATS

They look scary but most wētā are gentle animals and need protecting. They've been around for about 190 million years, but after the introduction of mammals such as cats and hedgehogs to New Zealand, wētā numbers plummeted.

INSECT ARMY

According to Māori legend, when the world began, wētā were part of an insect army recruited by Whiro, god of the underworld, to defeat his brother Tāne, god of forests. Tāne called on the winds and won the battle. He took all the insects down to Earth as prisoners and released them among the trees, where they still live now.

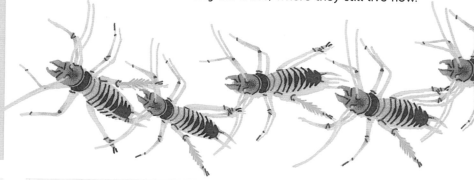

WĒTĀ MOTELS

Families across New Zealand are coming to the rescue by opening up wētā motels in their backyards. The motels give homeless wētā a safe place to hide and are a great way of observing these super-sized creepy-crawlies up close.

- The motels are made from hollowed-out blocks of wood with different rooms for the wētā to snuggle into. Some have a door so you can open it up to see what lives inside.

- Motels should be placed close to vegetation so the wētā don't have far to wander for dinner.

- The entrance needs to be narrow to keep out predators.

Spotting Wētā

WHERE TO SEE THEM

In wētā motels, trees, bushes and other damp, dark places.

WHEN TO SEE THEM

The best time to see them in action is at night, when they come out of their hidey holes to forage for leaves, seeds and flowers. Their favourite time for a wander is on warm, wet and very dark moonless nights.

LISTEN OUT FOR ...

'Tsit, tsit, tsit' — the sound they make by rubbing their legs against their bodies.

STINGLESS BEES
in Rio de Janeiro, Brazil

RIO IS FAMED FOR ITS TROPICAL BEACHES AND ITS COLOURFUL CARNIVAL, BUT IT HAS A WILD SIDE, TOO. Black vultures sunbathe on lamp posts, mischievous capuchin monkeys sneak into homes to steal fruit, and endangered species such as golden tamarinds, ocelots and sloths hang out in the huge forest that surrounds the city. Some of Rio's smallest animal inhabitants are also some of its most useful — stingless bees. Not only do they help to pollinate fruit and crops, they make delicious sweet honey!

Local beekeepers make homemade hives using old flowerpots, coconut shells, plastic bottles and other recycled containers.

In the suburb of Barra da Tijuca, there are bees living at the top of high-rise apartments!

Rio's beekeepers give bees a safe place to live. In return, the bees pollinate flowers and crops and make honey!

The makeshift bee homes are hung up inside houses, schools and apartments in the middle of the city.

Fuzzy body that helps collect pollen

Huge compound eyes

Tiny 'pollen baskets' on back legs used for carrying pollen back to the hive

Nectar is stored in a pouch inside the bee's stomach then regurgitated as honey back at the hive.

Rio de Janeiro is home to many species of honey-producing stingless bees (*Meliponini*) including the tiny 5mm-long jataí bee (shown here).
FAMILY Apidae (bees)
LENGTH 1.5–11 mm (0.06–0.4 in)
CONSERVATION STATUS Varies depending on species
WHERE IN THE WORLD? Central and South America, Africa, Southeast Asia and Australia

City Life

Bees are one of the most important pollinators in the world's forests and other wild spaces. But as more and more of these are destroyed, towns and cities can be an important refuge for bees. It's nice and warm in the city and there are lots of nooks to hide out in.

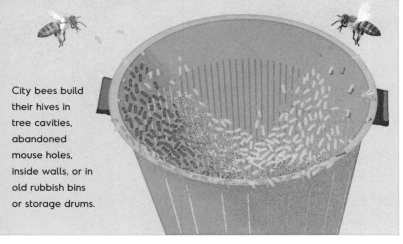

City bees build their hives in tree cavities, abandoned mouse holes, inside walls, or in old rubbish bins or storage drums.

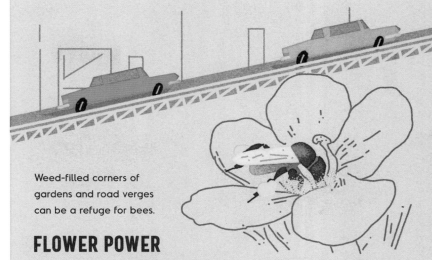

Weed-filled corners of gardens and road verges can be a refuge for bees.

FLOWER POWER

The loss of wild spaces and the use of pesticides on crops has led to declining bee populations around the world. Some species are in danger of dying out altogether. Towns and cities can be a rich source of food for these pollen-loving insects — thanks to the flowers that grow in nature reserves, gardens, allotments, cemeteries, parks, car parks, road verges and even on the roofs of high-rise buildings.

TINY FARMERS

Each time this busy little insect visits a flower, some of the dusty pollen clings to its hairy body, which it then transfers to other flowers. This helps the flower to make seeds from which new plants will grow. In fact, around a third of the foods we eat depends on bees and other pollinators!

Bees help all kinds of fruit and vegetables to grow, from apples, strawberries, pears and plums, to tomatoes, runner beans and onions.

SAVE THE BEES!

Bees need our help. By planting flowers and creating bee-friendly spaces near where we live, we can help native species of bees make a comeback. A bee hotel will give them a safe place to lay their eggs. You can make your own using an old plastic bottle and some hollow sticks.

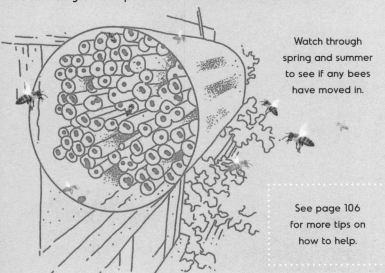

Watch through spring and summer to see if any bees have moved in.

See page 106 for more tips on how to help.

Spotting Bees

WHERE TO SEE THEM

Whizzing around flowery plants in gardens, parks, allotments and verges, or near bee hotels.

WHEN TO SEE THEM

On warm, sunny days as they forage for pollen.

LISTEN OUT FOR ...

Quiet buzzing sounds.

BUTTERFLIES
in Buenos Aires, Argentina

Butterflies can be as small as your fingernail or as large as your head.

ARGENTINA'S CAPITAL IS A JUMBLE OF OLD MANSIONS, GLASS SKYSCRAPERS AND GREY CONCRETE. But hidden in the urban maze are pockets of greenery, from balcony plant boxes to the amazing Costanera Sur Ecological Reserve — an oasis of wild grasses, trees and lagoons just minutes from the car-choked centre. In spring and summer, when the city's flowers are in bloom, the butterflies arrive to feed on nectar and lay their eggs on leafy plants.

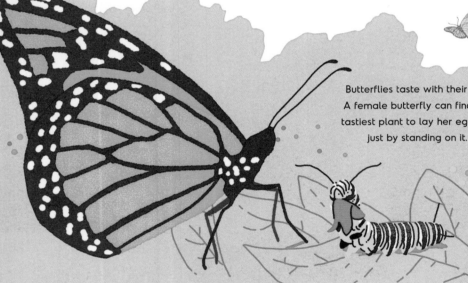

Butterflies taste with their feet. A female butterfly can find the tastiest plant to lay her eggs on just by standing on it.

Caterpillars have a ferocious appetite. They can grow as much as 100 times their original size.

There are lots of species of butterflies in Buenos Aires, including the southern monarch, mangrove buckeye, hortensia and passion butterfly (shown here).

SIZE 2 mm–28 cm (0.08–11 in)

CONSERVATION STATUS
Varies depending on species

WHERE IN THE WORLD?
Every continent except Antarctica

City Life

Plant-filled cities can be a real haven for butterflies and their caterpillars. Buenos Aires even has its own butterfly garden in the middle of the city, planted with butterfly-friendly plants. But as natural habitats are replaced with concrete and buildings, it can be tricky for butterflies to find enough food to eat or the right plants to lay their eggs on.

See page 106 to find out how you can help butterflies living near you.

THE BIG CHANGE

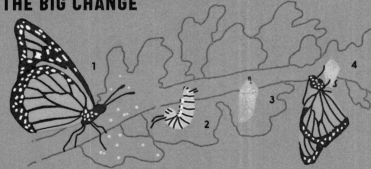

Butterflies are best known for the way they change from fat leaf-munching caterpillars to nectar-drinking butterflies. This amazing transformation is called metamorphosis.

1. First the female lays her eggs on a leafy plant.
2. When an egg hatches, a tiny caterpillar (known as a larva) crawls out and begins to chomp its way through its leafy surroundings.
3. Once it's big enough, it stops eating and forms a case around itself called a chrysalis or pupa.
4. After about 10 to 15 days it emerges as a beautiful butterfly.

Buenos Aires is famous for its football, tango and juicy steaks.

As they fly from plant to plant, butterflies spread pollen and help new flowers to grow. They help to pollinate all kinds of plants and crops.

Butterflies have a long straw-like mouthpart called a proboscis that they use to drink sugary nectar and juices.

HUNTING CATERPILLARS

The best places to look are under leaves with fresh nibble marks. Check around patches of long grass, in flower beds and hedges, or in the vegetable patch. For a closer look, put your hand or a piece of paper in front of its path and let it crawl aboard. Remember to return it to its leaf so it can carry on chomping.

Spotting Butterflies

TOP TIP: Beware of very hairy caterpillars — some can cause a rash if you touch them.

WHERE TO SEE THEM

Sheltering in roadside trees, flitting around flowers, or flying in flocks across the city. They can even be spotted in city centres and sometimes fly inside buildings.

WHEN TO SEE THEM

On sunny days in spring and summer.

LISTEN OUT FOR ...

The fluttering of their wings.

TOP TIP
Once you've spied a butterfly, stay very quiet and move slowly towards it (but don't get too close). It's better if you crouch down so you don't cast a shadow.

BIRDS

BIRDS IN THE CITY

OUR TOWNS AND CITIES ARE HOME TO ALL SORTS OF BIRDS. They can be spotted almost anywhere — soaring over buildings, perched in trees or bushes, pecking at crumbs outside cafes, or crowded around fountains and statues. Next time you go for a walk, or look out of the window, take a closer look!

Spotting Birds

Once you've spotted a bird, see if you can tell what kind of bird it is.
Here are some of the clues...

Size and shape

The size and shape of a bird are important clues to its identity. Take note of its body, beak, wings and tail. Here are a few examples of different bird shapes and sizes.

BROAD-BILLED HUMMINGBIRD

SUPERB STARLING

PEREGRINE FALCON

COMMON GULL

WHITE STORK

Shapes in the sky

You can sometimes identify a bird by looking at the shape of its wings and the way it's flying.

DIFFERENT TYPES OF WINGS

Some wings are designed for soaring, others for hovering, gliding, diving or flapping.

1. High-speed wings
Pointed, swept-back wings for speedy flying (e.g. falcons, swifts, swallows)

2. Fast, flappy wings
Perfect for taking off quickly to escape from predators and darting around trees and buildings (e.g. starlings, parrots, sparrows)

3. Broad, soaring wings
Perfect for soaring slowly and flying in circles (e.g. storks, vultures, owls, eagles, hawks)

4. Long, soaring wings
For gliding over long distances without too much flapping (e.g. gulls, albatrosses, gannets)

5. Hovering wings
Small and very quick for hovering in mid-air (e.g. hummingbirds)

FOOTPRINTS

Look out for tiny footprints in the mud or snow ...

Most bird tracks show three toes pointing forwards, and one back.

Water birds, such as ducks, have webbed feet for swimming and paddling.

Some birds, such as grebes, have bits of skin between their toes called lobes that get bigger and smaller as they swim.

Some woodpeckers, parrots and owls have two toes pointing forwards and two backwards. This helps them climb up trees without falling backwards.

RUNNING, JUMPING, SKIPPING

Some birds walk or run across the ground leaving a trail of single prints like this:

Others jump or skip so their tracks land near to one another, like this:

Nests in the City

Some urban birds nest in trees and bushes. Others snuggle up inside churches, under bridges, on window ledges or balconies and on the rooftops.

Some birds make their homes in more unusual places. Robin nests have been found in tin cans, abandoned cars and rubbish bins. One was even discovered under the lid of a barbecue!

REMEMBER

If you discover a bird's nest it's very important not to touch it or to disturb the birds inside.

FEEDING SIGNS

Nibbled nuts, fruit and seeds are another sign you can look out for.

Thrushes use stones and hard surfaces to break open snail shells. Look out for small piles of broken shells on pavements, stone steps or concrete drain covers.

TOP TIPS

- Gardens are a great place to see city birds. You can attract even more birds by providing a bird feeder or a nesting box.

- Listen out for the songs they sing.

- Be still and quiet (you don't want to scare them away!).

Finding Feathers

Every bird has thousands of feathers covering its body. They help them to fly, keep them warm and protect them from the wind, rain and sun. But all that flapping, as well as knocks and scrapes against trees, lamp posts and buildings, can cause wear and tear. So, every now and then birds shed their feathers and grow new ones. This is called moulting. Discarded feathers are one of the most common signs that a bird leaves behind.

WHICH FEATHER?

Birds have different feathers covering different parts of their body. They each have their own special job to do.

1

1. DOWN FEATHERS

Down feathers are short and fluffy. They sit under the contour feathers for extra insulation. They're so soft and warm that people use them to fill duvet covers and pillows.

2. CONTOUR FEATHERS

Contour feathers are smaller than flight feathers. They lie flat against a bird's body making it streamlined for flying and keeping it warm and dry.

3

2

3. FLIGHT FEATHERS

These can be found on the wings or the tail. They are the longest feathers and are very strong as they help to propel a bird forward in flight. Most tail feathers are also used for steering, just like a rudder.

FEATHER CARE

Just as you spend time washing and brushing your hair, birds spend time tidying and cleaning their feathers. This is called preening. Birds preen for lots of reasons:

1. To wipe away dirt and parasites
2. To rearrange their feathers so they're in the right position for flying
3. To waterproof their feathers. Most birds have a gland near their tail that produces an oil-like substance. They use their bill, head or feet to rub this oil over their feathers to keep them waterproof.

Some species can have as many as 25,000 feathers, so preening can take up a lot of their time. Next time you see a bird playing with its feathers you'll know what it's up to!

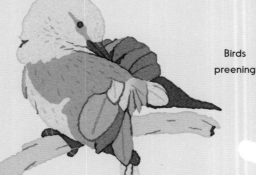

Birds preening

Which Bird?

Sometimes you can tell what bird the feather belongs to by looking at its shape, colour and markings. The different colours and patterns can help a bird to hide from predators or to attract a mate. Some species have fancy head or tail feathers to show off or to scare away intruders.

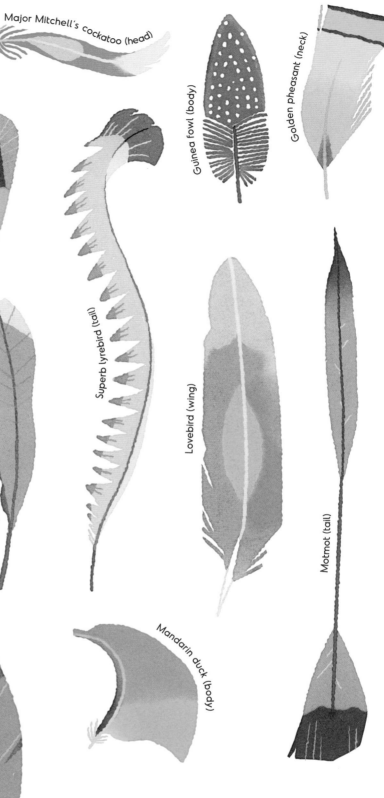
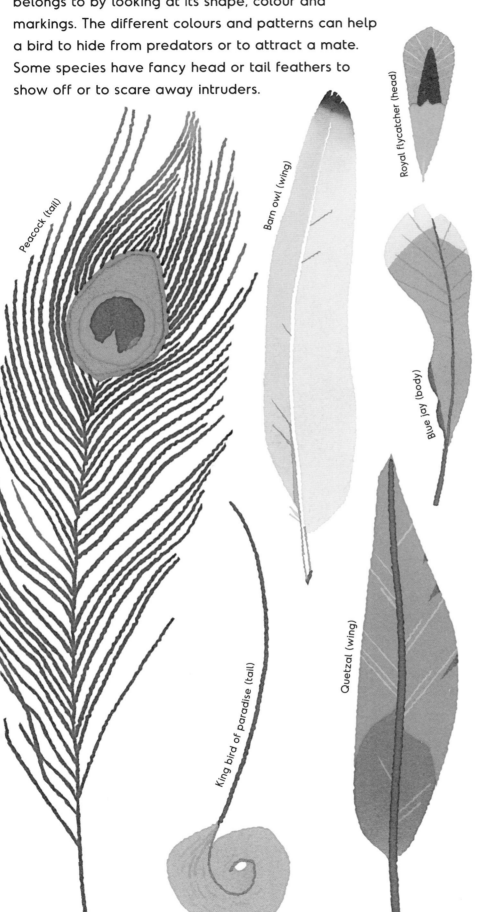

Major Mitchell's cockatoo (head)

Golden pheasant (neck)

Guinea fowl (body)

Royal flycatcher (head)

Barn owl (wing)

Superb lyrebird (tail)

Peacock (tail)

Blue jay (body)

Lovebird (wing)

Motmot (tail)

King bird of paradise (tail)

Quetzal (wing)

Mandarin duck (body)

SPOTTING CHALLENGE

If you find a feather lying on the ground, look at the colours and patterns and see if you can work out what bird it belongs to. Then see if you can tell which kind of feather it is. Is it a flight feather or a contour feather? Or a fluffy down feather?

TOP TIP

Flight feathers are usually asymmetrical — meaning one side is a bit bigger than the other.

PEREGRINE FALCONS

in Barcelona, Spain

SOARING OVER THE ROOFTOPS OF BARCELONA, THE SAGRADA FAMILIA CHURCH IS ONE OF THE CITY'S MOST SPECTACULAR SIGHTS. If you look up at its ornate spires you might see a family of peregrine falcons that have made their home in the bell tower. While the young chicks chatter in the nest, the mother perches on a ledge, scouring the land below for her evening meal. Spotting a plump pigeon, she swoops down and speeds towards her prey ...

The Sagrada Familia was designed by the famous architect Antoni Gaudí

When viewed from below, a peregrine's pale belly is hard to see against the sky and clouds. This is the perfect camouflage for a hunter chasing its prey.

Pointed, swept-back wings and streamlined body

Sharp bill to bite the neck of its prey

On the streets below, groups of sightseers explore the city by bus.

Adults have blue-grey upper parts that blend in with the ground below, making them tricky to see from above.

LATIN NAME *Falco peregrinus* (meaning 'wandering falcon' — it got its name because it loves to travel)
FAMILY Falconidae (falcons)
LENGTH 39—50 cm (15—20 in)
WINGSPAN 95—115 cm (3—3.7 ft)
CONSERVATION STATUS Least Concern
WHERE IN THE WORLD? In most parts of the world, except in tropical rainforests and polar regions

Its eyesight is so sharp it can spot its prey more than 1.6 km (1 mile) away.

Pale throat and cheeks and dark 'moustache' stripes

Powerful feet with black talons like daggers

City Life

Peregrine falcons have set up home in cities around the world. The city reminds them of their natural habitat where they build nests on the tops of cliffs or quarries. Tall buildings are perfect perches for scanning the local hunting area and raising their young. There's plenty to eat too — all kinds of smaller birds live in cities or fly through them as they migrate.

CITY LIGHTS

Peregrines also take advantage of bright city lights to hunt during the night. In Barcelona, artificial lights directed at buildings help peregrines to see bats and songbirds that usually fly under the cover of darkness. In New York, they hunt by the glow of the iconic Empire State Building.

WHERE DO THEY LIVE?

Power stations, phone masts, museum rooftops, office blocks, churches or anywhere up high with a ledge to perch on. Peregrines have been found nesting near some of the world's most famous high places ...

Perched on the Houses of Parliament in London, United Kingom

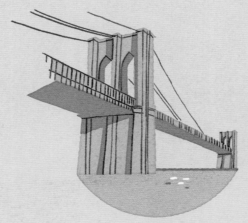

Inside the towers of New York's busy Brooklyn Bridge, USA

ROOM WITH A VIEW

A falcon's nest is called a scrape. They have some of the best views over the city!

PERFECT HUNTER

The peregrine falcon is built for speed. It has a long, slender body and its wings are pointed and swept back — like the wings of a jet plane. This reduces wind resistance and allows the falcon to slice through the air at record-breaking speeds. When diving, it can reach 300 kph (186 mph), making it not only the fastest bird, but also the fastest animal in the world.

1. Once it spies its target, it turns, dives and accelerates by beating its wings.

2. It then tucks in its wings and plummets through the air like a bullet. This movement is known as a stoop.

3. As it gets close, it opens its wings and snatches its prey in mid-air.

WHAT DO THEY EAT?

Peregrine falcons feed mainly on birds, and nearly all birds found in urban parks are on the menu. One of their favourite foods is pigeon — and cities are full of them. If you keep your eyes open you may spot them hunting!

FALCONS TO THE RESCUE!

In Rome, peregrine falcons have been brought in to scare away the seagulls and blackbirds that have been pecking at the Colosseum — the famous ancient arena where gladiator contests once took place. Falcons are also used near airports to stop smaller birds from bumping into plane windows or flying into the engines.

Spotting Falcons

SAVING FALCONS

Populations of peregrines around the world were once in danger of extinction due to the widespread use of a pesticide called DDT. The banning of DDT allowed peregrine populations to recover. This is one of the planet's biggest conservation success stories.

TOP TIP

Remember to look up. You may spy them flying in the sky or sitting near their high-rise nests.

WHERE TO SEE THEM

The areas around the nest are the best place to watch them as adults fly back and forth to feed their young. Pigeon feathers or pigeon remains around the bottom of buildings are one of the clues that a nest is nearby.

WHEN TO SEE THEM

Like many birds, peregrines are diurnal, which means that they are active mostly during the day. They're more visible in the breeding season. This varies across the planet but is usually during the spring and summer months. In Barcelona the best time to see them is between February and September.

LISTEN OUT FOR...

'Kak-kak-kak' — a harsh alarm call that they use when danger is near

'Klee-klutch' — the musical-sounding call of mating pairs

'Chup-chup' — the noise made by chicks as they're being fed

ALPINE SWIFTS
in Naples, Italy

Long, curved wings and short forked tail — the perfect aerodynamic shape

PERCHED ON ITALY'S WEST COAST, THE BEAUTIFUL CITY OF NAPLES LIES ON THE MIGRATION PATH OF ONE OF THE WORLD'S FASTEST AND MOST WELL-TRAVELLED BIRDS — the Alpine swift. Each summer, they fly across continents to return to their old nesting sites. High above the rooftops, they dive and turn like tiny acrobats ... before swooping into hidden holes and cracks in the city's ancient buildings.

Dark, sooty colour with a white belly and throat

An individual swift might fly more than a million kilometres (620,000 miles) in lifetime — the same as travelling 25 times around the Earth.

Swifts only come down from the air to lay their eggs and raise their chicks. Some fly for as long as ten months without landing once!

Naples is one of the oldest cities in Europe.

LATIN NAME
Tachymarptis melba
FAMILY Apodidae (swifts)
LENGTH 20—23 cm (8—9 in)
WINGSPAN 53—60 cm
(21—24 in)
CONSERVATION STATUS
Least Concern
WHERE IN THE WORLD?
In summer Europe and Asia where they breed and raise their chicks, in winter Africa

City Life

Traditionally, swifts have nested in small holes in cliffs and trees, but now they mainly nest in buildings — under loose tiles or in gaps in walls, in eaves, lofts, church spires and towers. They're more likely to be seen in older parts of towns as new buildings don't have as many nooks and crevices for them to snuggle into.

In Naples, Alpine swifts nest in holes in the walls of the National Archaeological Museum.

NEW HOMES

Swifts return to the same nests year after year, but as old buildings are repaired or demolished, many swifts are left homeless. Wooden nest boxes can give visiting swifts a brand-new place to stay.

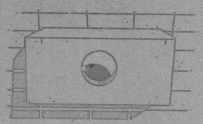

CHIMNEY SWIFTS

Every evening from mid-August to mid-October, thousands of Vaux's swifts gather in the sky above Chapman Elementary School in Portland, USA. Then they zoom down the chimney, like long plumes of smoke in reverse, to sleep for the night!

Migrating swifts often use chimneys as roosts (places to sleep) as they prepare for their long journey south.

HOLY SWIFTS

In the holy city of Jerusalem, around 88 pairs of common swifts return each year to their nests inside the ancient Western Wall. Plans to cement up holes in the wall were abandoned so that the swifts' tiny homes could remain intact.

The Western Wall is an important religious site for Jewish people.

LIFE ON THE WING

Swifts spend most of their life 'on the wing'. This means that they eat, drink, mate and even sleep in the air! As they dart through the sky, they grab flying insects and spiders. They drink by swooping low over water or catching raindrops. To rest, first they fly really high, then they slowly glide down, taking short naps as they go.

Swifts feed their young with 'food balls' that they collect in a pouch at the back of their throat. They can store as many as 600 insects in a single ball!

Spotting Swifts

Swifts have a very distinctive shape when flying — they look a bit like boomerangs!

WHERE TO SEE THEM

Flying fast and low around buildings or swooping into their nest holes.

WHEN TO SEE THEM

Early in the morning, or late at night as they fly home to roost.

LISTEN OUT FOR ...

Loud, high-pitched screaming as groups of swifts fly overhead.

GREAT CRESTED GREBES
in Amsterdam, The Netherlands

The canals were built in the 17th century to transport goods and people around the city.

This bird has got tangled up in some plastic netting. Swallowing bits of plastic that have been dropped in the water can also be deadly.

Reddish head crest and ornate head plumes

City Life

Amsterdam is famous for the canals that wind through the city. There are 165 in total, running to a length of more than 100 km (60 miles). All kinds of plants and fish live in the water, as well as crabs and crayfish . . . and this attracts aquatic birds such as herons, swans, geese and grebes!

DANCING BIRDS

The spectacular dance of the great crested grebe is performed by courting couples just before they build their nests.

LATIN NAME
Podiceps cristatus
FAMILY Podicipedidae (grebes)
LENGTH 46–51 cm (1.5–1.7 ft)
WINGSPAN 85–90 cm (2.8–3 ft)
CONSERVATION STATUS
Least Concern
WHERE IN THE WORLD?
Much of Europe and Asia and parts of Africa, Australia and New Zealand

First, the grebes greet each other by shaking their heads and showing off their head crests.

Next, they dive under the water to collect weeds and twigs in their beaks.

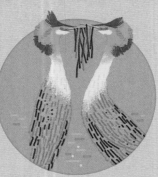
Finally, they rise up, chests together, while quickly paddling their feet to keep themselves afloat.

WITH MILES OF CANALS TEEMING WITH FISH, AMSTERDAM IS A WATERY WONDERLAND FOR BIRDS. Herons stand on long legs at the water's edge, swans glide gracefully by, and flocks of noisy geese skim overhead. In springtime, pairs of great crested grebes can be spotted strutting their stuff as they perform their curious courtship dance, before building floating nests of twigs and rubbish.

The people of Amsterdam love to cycle!

Sharp pointy beak

Like many city birds, grebes sometimes use bits of rubbish floating in the water to help build their nests.

Young chicks hitch a ride on the back of one parent while the other feeds them small fish.

Baby grebes have black and white stripes and look like mini zebras!

PLASTIC WHALES

A company called Plastic Whale has come up with a clever plan to tackle the city's plastic pollution: plastic fishing! Tourists are taken on a sightseeing boat tour along the historic canals, then help to scoop up rubbish using fishing nets.

The plastic isn't thrown away — it's used to make new boats!

PLASTIC ISLANDS

Plastic does have its uses. In Glasgow, UK, special floating islands made of old plastic and planted with vegetation give young grebes a safe place to grow up.

Grebe nests can be washed away by heavy rain, putting the chicks in danger. Plastic islands are a safe and secure nesting place.

Spotting Grebes

WHERE TO SEE THEM

Grebes are not very good flyers, so you're more likely to see them drifting along canals and rivers than flying in the sky.

WHEN TO SEE THEM

Spring and summer are the best times to see the dancing birds and fluffy new chicks.

LISTEN OUT FOR …

'Vrek-vrek-vrek' — the mating call of great crested grebes
'Pli-pli-pli' — the noise of young grebes begging for food

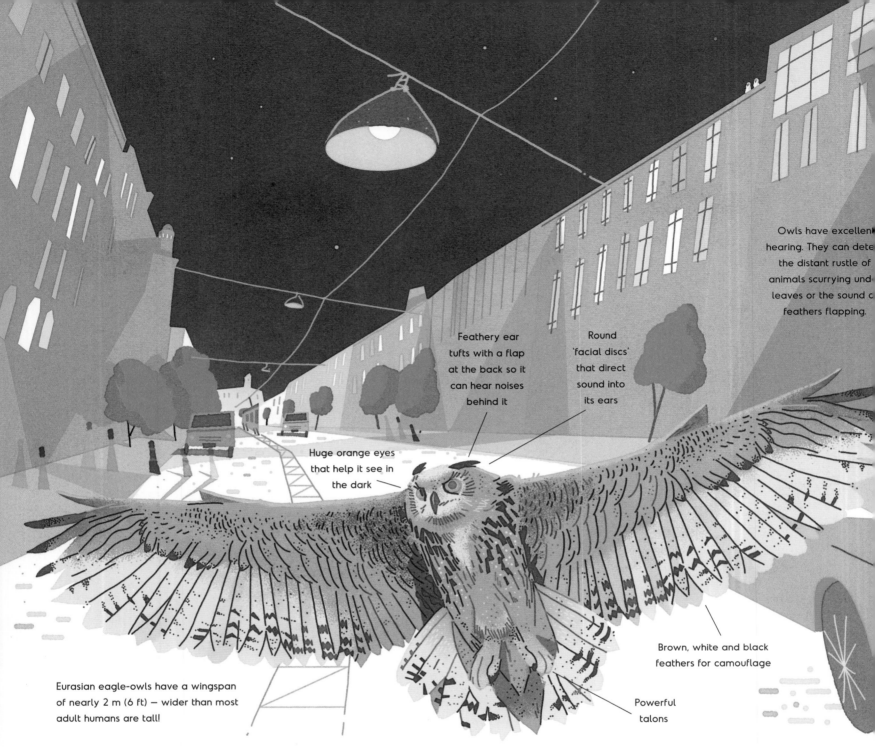

Owls have excellent hearing. They can detect the distant rustle of animals scurrying under leaves or the sound of feathers flapping.

Feathery ear tufts with a flap at the back so it can hear noises behind it

Round 'facial discs' that direct sound into its ears

Huge orange eyes that help it see in the dark

Brown, white and black feathers for camouflage

Powerful talons

Eurasian eagle-owls have a wingspan of nearly 2 m (6 ft) — wider than most adult humans are tall!

EURASIAN EAGLE-OWLS

in Helsinki, Finland

Eagle-owls can even hear animals moving around under snow.

IT'S RUSH HOUR IN DOWNTOWN HELSINKI. Trams jingle their bells as crowds of people hurry home from work. High above the hubbub, perched on a rooftop, a family of Eurasian eagle-owls peeps down on the scene below. While the mother owl stands guard over the chicks, the father spreads his massive wings and swoops silently over the city to hunt wild rabbits and hares…

LATIN NAME *Bubo bubo*
FAMILY Strigidae (true owls)
LENGTH 56–75 cm (1.8–2.5 ft)
WINGSPAN 1.4–1.8 m (4.6–5.9 ft)
CONSERVATION STATUS
Least Concern
WHERE IN THE WORLD?
Europe and Asia

City Life

In the wild, eagle-owls live on rocky cliffs or in forests, but there are lots of perks to city life. Buildings make great hunting and nesting perches where owls can watch and listen for prey, and there are all kinds of small animals to eat. In Helsinki, the parks are brimming with wild rabbits and hares.

One family has made its home on top of a shopping centre opposite the main train station.

Birdwatchers flock to the roof of a nearby hotel to get a sneaky peek at the owls.

Owls don't make loud flapping noises like other birds do. Their soft feathers help to muffle sound, so they can swoop down on their prey in complete silence.

FLYING TIGERS

Eagle-owls are sometimes known as 'flying tigers' due to their huge size and reputation as fierce predators that will prey on animals as big as foxes! First, they grab their victim with their talons, then they break its neck or crush it to death. Smaller animals are eaten whole, while bigger creatures are torn into pieces first.

BIRD-SCARERS

Specially trained eagle-owls are used to scare away troublesome city seagulls that leave behind droppings and snatch food from smaller birds (and people)!

An eagle-owl known as 'Yoda' has been helping to keep the University of Bath, UK, clear of pesky seagulls.

CITIZEN OF THE YEAR

A football-loving eagle-owl staged a surprise pitch invasion at the Helsinki stadium right in the middle of a Euro 2008 qualifier match. Finland went on to win the game, and the owl — nicknamed 'Bubi' — was named Citizen of the Year!

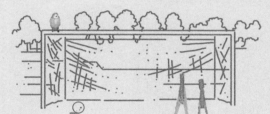

The Finnish national football team love their lucky Bubi so much that they are now known as Huuhkajat (Finnish for Eurasian eagle-owl)!

Spotting Owls

WHERE TO SEE THEM

Roosting on buildings, poles and church towers. In parks and cemeteries and other green areas of city centres.

WHEN TO SEE THEM

Resting on high-up perches in the day. Hunting over the city at night.

LISTEN OUT FOR ...

Owls hooting in the dark. Calls range from a deep, booming 'Ooh-hu' to high-pitched, bark-like screams. They're most chatty on moonlit nights!

Look out for owl pellets near nesting spots.

The parts of a prey animal that the owl can't digest, such as bones, fur and feathers, are regurgitated as pellets. By looking at the pellets you can work out what the owl has had for its dinner!

WHITE STORKS
in Marrakesh, Morocco

WALKING THROUGH THE MEDIEVAL CITY OF MARRAKESH IS like stepping back in time. A maze of alleyways lead to treasure-filled souks and magnificent mosques dot the sky. Right here, in the middle of the old town, long-legged white storks strut along crumbling fortress walls and build big stick nests on rooftops and towers!

White feathers on its head, neck and body

Black feathers on the edge of its wings

Koutoubia mosque is the tallest mosque in the city.

Long, pointed red beak

White storks sometimes share their nests with smaller birds such as sparrows.

The nests are made with sticks and leaves and are HUGE — up to 1.8 m (6 ft) wide and 2.8 m (9 ft) deep!

Long, broad wings
for soaring
on thermals

LATIN NAME *Ciconia ciconia*
FAMILY Ciconiidae (storks)
LENGTH 1–1.15 m (3.2–3.9 ft)
WINGSPAN 1.5–2.2 m (4.9–7.2 ft)
CONSERVATION STATUS Least Concern
WHERE IN THE WORLD? Spring and summer breeding
and raising young in parts of Europe, North Africa, Asia
Minor and the Middle East; winter in tropical Africa,
parts of the Middle East and the Indian subcontinent

Long,
slender
neck

Long legs
for wading

City Life

White storks usually build their nests on top of tall trees, but they're also happy to use buildings and other man-made structures. They like high-up places so they can keep their eggs and chicks safe from predators.

In Marrakesh, there are lots of towers, chimneys and rooftops for storks to nest on.

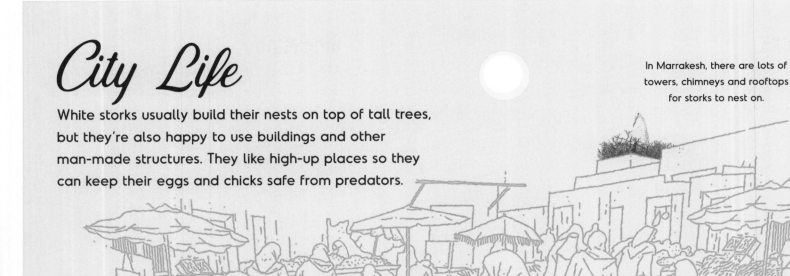

STORKS OF THE WORLD

White storks nest on top of electric pylons in Andalucia, Spain.

One-and-a-half metre (5 ft)-tall jabiru storks stroll down the streets in Corumbá, Brazil.

Marabou storks in Nairobi, Kenya, have been upsetting local residents by showering passers-by with droppings. They build their nests in trees next to the national football stadium and sometimes strut across the pitch!

There's no shortage of lakes in Lakeland, Florida, USA. Wading birds love it here. These wood storks have even been known to steal food from young alligators at a nearby alligator park!

HIGH-RISE LIVING

Nests on power poles can disrupt electricity supplies and are dangerous for the birds. Artificial platforms placed above the power cables give the birds a safer place to build their homes.

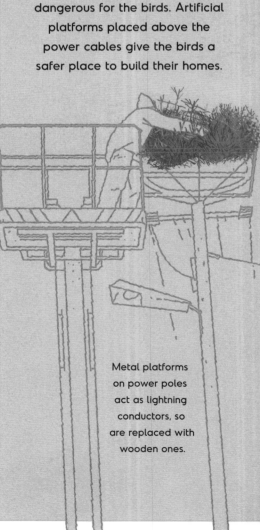

Metal platforms on power poles act as lightning conductors, so are replaced with wooden ones.

TEAMWORK

White storks build their nests in pairs. The males collect most of the materials, and the females do most of the building.

JUNK FOOD

Thousands of white storks in Spain and Portugal have set up home close to giant rubbish dumps, where they line their nests with junk, and snack on leftover burgers and sandwiches. Some have become so used to the quick-and-easy junk-food meals that they stay here all year round instead of migrating south in the winter months.

Staying near the dumps may spare the birds a long and treacherous journey to their usual winter homes, but a diet including plastic, metal and rubber bands can be deadly.

MAGICAL BIRDS

According to an old Berber legend, storks are actually human beings who magically transform themselves into graceful, long-legged birds when they want to travel.

Before people knew about bird migration, this story helped to explain the mysterious disappearance of storks in the winter.

In European folklore the stork is responsible for delivering babies to their families.

Spotting Storks

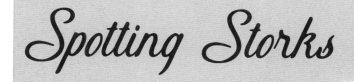

WHERE TO SEE THEM

Watch out for storks flying overhead with their long neck and legs outstretched. Check telegraph poles, rooftops and chimneys for their nests — they're big, so are easy to spot!

WHEN TO SEE THEM

During the day as they build their nests and hunt.

LISTEN OUT FOR ...

'Clack-clack-clack' — the sound of loud clattering as storks open and close their beaks to greet one another. Young storks mew like a cat when they beg for food!

AFRICAN PENGUINS

in Simon's Town, South Africa

Streamlined body and long flat wings

Penguins can't fly, but they are champion swimmers. They use their wings as flippers to 'fly' through the water

Penguins can swim as far as 70 km (43 miles) in a single trip, and some species can reach speeds of more than 35 kph (22 mph)!

Short tail and feet to help it steer

Their eyesight is specially adapted for seeing in the dark depths of the sea — and helps them to find their way around town at night!

White tummies to keep them hidden from underwater predators looking up and black backs to hide from predators looking down from above

Body covered in very short, stiff feathers, forming a cosy, waterproof swimming costume

BEFORE CITIES DEVELOPED ALONG SOUTH AFRICA'S COAST, mainland beaches were a wild and dangerous place for animals like the African penguin. With their awkward waddle, they would have quickly been devoured by leopards and other predators that prowled nearby. But as towns grew, the predators began to move elsewhere — and so, in moved the penguins!

LATIN NAME *Spheniscus demersus*
FAMILY Spheniscidae (penguins)
HEIGHT 60–70 cm (2–2.3 ft)
CONSERVATION STATUS
Endangered
WHERE IN THE WORLD?
Southwest Africa

City Life

In the 1980s a pair of African penguins was spotted nesting on Boulders Beach in Simon's Town, a small fishing community near Cape Town. More followed, and today there is a colony of around 2000 pairs. The penguins have grown used to people and go about their business as visitors watch from wooden boardwalks.

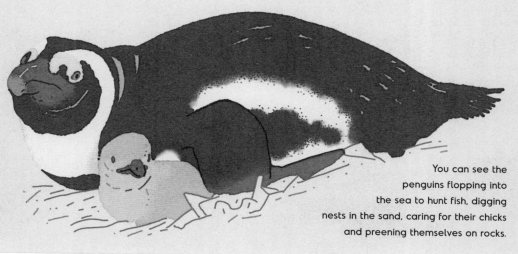

You can see the penguins flopping into the sea to hunt fish, digging nests in the sand, caring for their chicks and preening themselves on rocks.

PENGUIN RULES

1. DON'T disturb the penguin nesting sites (stay on the boardwalks).
2. DON'T touch, chase or feed the penguins.
3. BEWARE! African penguins have razor-sharp beaks and can give you a nasty bite if you get too close and they feel threatened.

OUT AND ABOUT

Fences have been put up to stop the penguins from wandering away from the beach but they have been known to waddle into some unexpected places!

Raiding nearby gardens for nest material and sneaking inside houses

Trotting along the streets of Simon's Town in search of their nests

Napping under parked cars and sheltering in drains or roadside vegetation

SAVING PENGUINS

These little birds need our help! Due to oil spills, egg poaching, over-fishing and climate change, some species are now critically endangered.

In Simon's Town, wooden nest boxes have been installed on the beach to help shelter the penguins from the sun and protect their eggs from hungry gulls.

In the town of Timaru, New Zealand, a mini underpass has been built to help little blue penguins cross busy roads between their nests and the sea.

Spotting Penguins

WHERE TO SEE THEM

On city beaches close to penguin feeding grounds and nesting sites. You can smell them, too! Thanks to their diet of sardines and anchovies they whiff of fish.

WHEN TO SEE THEM

Early morning as the penguins set out on their fishing trip or late afternoon when they return.

LISTEN OUT FOR...

A loud 'braying' noise that sounds like a donkey — the mating call of the African penguin!

MANDARIN DUCKS
in Beijing, China

Mandarin ducks usually pair for life. In Asia they are thought to be a symbol of love and good luck.

Mandarin ducks feed mainly on insects, snails, fish, plants and seeds.

The male Mandarin duck uses its bright feathers to impress females. Females aren't as colourful as the males.

If you see a male with its neck outstretched and its crest and orange 'sails' raised, this means it's flirting with the lady ducks!

They find their food by dipping their bills into the water, upending (feeding upside down in the water) or grazing on land.

Their webbed feet act like paddles and waterproof feathers keep them dry and warm.

BEIJING IS ONE OF THE FASTEST-GROWING CITIES ON THE PLANET. WILDLIFE IS THE LAST THING YOU'D EXPECT TO SEE HERE, but tucked away in this concrete forest are hundreds of species of animals. Weasels and chipmunks scurry through alleyways looking for snacks, tiny swifts build nests in the imperial palace walls, and paddling about in the city's lakes and rivers is one of the world's most beautiful and graceful birds – the Mandarin duck.

LATIN NAME *Aix galericulata*
FAMILY Anatidae (ducks, geese and swans)
LENGTH 41–49 cm (16–19 in)
WINGSPAN 68–74 cm (27–29 in)
CONSERVATION STATUS Least Concern
WHERE IN THE WORLD? Native to China, Japan, Korea and parts of Russia, introduced to parts of Europe and the USA

City Life

It's no secret that ducks love water — and many cities have lots of ponds, lakes, marshes and rivers for ducks to paddle in. In Beijing, China's past emperors built beautiful green pleasure gardens with ornate temples and pagodas (tiered towers), alongside huge lakes. Today they're popular with both tourists and ducks!

TREE HOUSES

Mandarin ducks nest in holes in trees close to the water. Just one day after hatching, the brave young ducklings dive out of their high-rise apartment, one by one, to the ground below. But don't worry, they bounce! The mother then leads them to the water for their first swim.

They're so light and fluffy the fall doesn't hurt them.

ESCAPE ARTISTS

These adventurous ducks are expert escape artists and often break out from zoos and private collections. Others have been set free by their owners. Because of this, they can now be found in cities far away from their native habitats, including New York City and London.

CITY DUCKS

Mandarin ducks aren't the only water-loving birds to make a go of it in the big city. Other species of ducks are a common sight in urban places around the world.

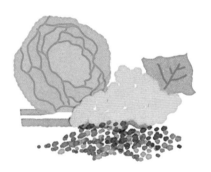

You might find them nesting in some strange places — gardens, balconies, rooftop terraces, even concrete islands in the middle of car parks!

In cities, ducks will make use of whatever watery places they can find, like this fountain.

Look for them sipping rainwater from roadside puddles ...

...or taking a dip in a swimming pool.

FEEDING DUCKS

Feeding ducks processed food like bread can actually be harmful to them. Next time you go to the park, make sure you take natural nibbles such as corn, seeds, lettuce or grapes with you.

Spotting Ducks

WHERE TO SEE THEM

In ponds, lakes, rivers and streams, drinking from puddles, sneaking into swimming pools, perched in trees or nesting in tree hollows.

WHEN TO SEE THEM

Early morning and late afternoon, when they're most active.

LISTEN OUT FOR ...

Quacks, coos, whistles and grunts. Female mandarins make clucking sounds! If you hear this, it could be a warning that danger is nearby.

The Royal Botanic Garden is a popular parrot hangout.

Short, sharp beaks and long tongues for reaching into flowers

Rainbow lorikeets are very sociable birds and gather in big flocks to feed.

Long claws, and two toes pointing forward and two pointing backward — perfect for grasping perches and gripping food while passing it up to their bill

Like other parrots, rainbow lorikeets are excellent mimics and can imitate all kinds of everyday sounds, including telephones, car alarms and microwave pings!

Pointed wings and tail help it fly super-fast

LATIN NAME *Trichoglossus moluccanus*
FAMILY Psittaculidae (parrots)
LENGTH 25–30 cm (9.8–11.8 in)
WINGSPAN 45–46 cm (17.7–18 in)
CONSERVATION STATUS
Least Concern
WHERE IN THE WORLD? Parts of Australia

RAINBOW LORIKEETS
in Sydney, Australia

WILDLIFE IS NEVER FAR AWAY IN SYDNEY! Giant bats hang out in suburban fig trees, possums sneak into attics and oversized spiders and cockroaches scuttle about under porches. Among the city's most popular feathery residents is the clown-like rainbow lorikeet. Early each morning, as boats ferry people across the harbour, these colourful parrots screech above the famous Opera House in noisy squadrons before congregating in trees to lap up nectar and pollen.

Visitors love sunny Sydney for its sandy surf beaches and friendly locals.

The harbour is brimming with ocean creatures — from crustaceans and jellyfish to dolphins, whales and sharks!

Parks and gardens cut through the skyscrapers and the whole city is ringed by national parks.

City Life

Rainbow lorikeets can be seen in towns and cities all along the east coast of Australia. They feed mainly on fruit, pollen and nectar, and Sydney is full of flowering trees for them to dine on. They are also nifty fliers and can duck and weave around buildings and other obstacles with ease.

RAINBOW CAMOUFLAGE

They may be colourful, but rainbow lorikeets are surprisingly good at concealing themselves. The green feathers on their back and tail help them hide in leafy trees. Their blue head resembles the sky and their bright tummy and beak look like flowers. Only their noisy chatter gives them away!

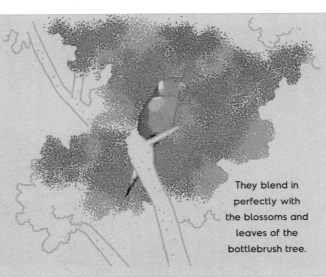

They blend in perfectly with the blossoms and leaves of the bottlebrush tree.

TONGUE BRUSHES

The rainbow lorikeet has an ingenious trick for scooping up nectar and pollen — the tip of its extra-long tongue is shaped just like a bristle brush!

DAY IN THE LIFE OF A SYDNEY RAINBOW LORIKEET

Just like Sydney's commuters, the city's lorikeets have a schedule to keep!

The day begins with a commute across the city from their roosts to their feeding grounds. Early in the morning, when it's cooler, is their busiest time.

As the day heats up, they may stop to rest or socialise for a while. They can often be heard squabbling in trees or seen cooling off in bird baths.

It's important to keep hydrated after all that flapping about in the Australian sun, so they stop off at water points throughout the day.

In the afternoon they'll feed some more, before flying home with the rest of the crowds at sunset.

After jostling for the best perch, it's lights out in their communal roosts. Tall eucalyptus trees with hollows in the trunks make the best nests.

PARROTS ON THE RUN!

Rose-ringed parakeets are probably the most widely distributed parrots in the world. Native to southern Asia and central Africa, they can now be found in cities in Europe, USA, South Africa, Egypt and the Middle East. Many are descended from pets that have escaped or been set free.

In their native India, rose-ringed parakeets use their sharp bills to rip open sacks of grain at railway depots.

PARROT PARADISE

Rainbow lorikeets aren't the only parrots in Sydney. Pretty pink-chested galahs hang out in suburban trees and raucous sulphur-crested cockatoos can often be heard screeching their heads off early in the morning.

City-slicker cockatoos have learnt how to open wheelie bins and break in through kitchen windows to search for nuts.

Sulphur-crested cockatoo

Pink-chested galah

PARROTS OF BROOKLYN

Wild monk parakeets build nests the size of small cars on the streets of Brooklyn in New York City, USA. The birds are native to South America, but it's thought that their descendants slipped out of containers at New York's JFK Airport in the 1970s and made a break for freedom!

They've even been seen eating slices of pizza in the trees!

Spotting Parrots

WHERE TO SEE THEM
Gardens and parks with tall trees and flowering plants.

WHEN TO SEE THEM
Early mornings as they leave their nests. Or late afternoon, when they head home.

LISTEN OUT FOR...
Very loud screeching and high-pitched whistles and chattering.

BROAD-BILLED HUMMINGBIRDS
in Mexico City, Mexico

Hummingbirds' wings move so fast they make a humming sound. Some species can beat their wings as many as 80 times a second!

They use their specia[l] designed tongues to l[ap] up nectar at about [20] licks per second.

Hummingbird feathers are iridescent, which means that they change colour depending on how the sunlight hits them.

They can hover, dive and fly upside down and backwards ... but they can't hop or walk!

Hummingbirds are very important pollinators. They help new plants to grow.

Mexico City was built on the site on the ancient city of Tenochtitlán and was once home to powerful Aztec warriors.

ONCE AT THE HEART OF THE MIGHTY ANCIENT AZTEC EMPIRE, Mexico City is the world's second-biggest urban centre — and it's still growing. But even here, there's room for wildlife. Thousands of species of animals can be found in the patches of forest and green spaces dotted across the city. Among them, zooming about like tiny jewelled helicopters, are some of the smallest, fastest and most dazzling birds on the planet — hummingbirds!

LATIN NAME *Cynanthus latirostris*
FAMILY Trochilidae (hummingbirds)
LENGTH 8–10 cm (3–3.9 in)
CONSERVATION STATUS
Least Concern
WHERE IN THE WORLD?
Southern USA and Mexico

City Life

Mexico City is home to more than a dozen species of hummingbirds. Over the years, much of their natural forest habitat has been cut down and replaced by buildings. But by hanging up nectar feeders and planting nectar-rich flowers in backyards, rooftop gardens and window boxes, the local people are helping to save the city's hummingbirds.

CITY HOMES

Hummingbirds weave cup-shaped nests from spider webs, plant fibres, twigs and leaves. Urban birds have found some unusual places to perch their tiny homes, from washing lines and TV cables to wind chimes and Christmas lights!

New York City's parks and gardens are a popular pit-stop for ruby-throated hummingbirds.

EPIC JOURNEYS

They may be tiny, but hummingbirds are intrepid travellers. Some species migrate hundreds of miles across North America to their winter homes in Central America. It's a long, gruelling journey so city gardens can be their lifelines. They give them a place to stop and rest a while and to refuel on sugary plants.

MYTHS AND LEGENDS

Many myths and stories have been told about the amazing hummingbird.

The Aztecs believed that warriors killed in battle were transformed into hummingbirds. Their god of war, Huitzilopochtli, was also depicted as one.

It was once thought that hummingbirds migrated south from Canada by hitching a ride on the back of geese!

HUMMINGBIRD GARDENS

If you have hummingbirds in your area, here's how you can help.

1. Try planting hummingbird-friendly flowers in your garden, or in hanging baskets outside your window. Red tubular flowers that hold lots of nectar are their favourites!

2. A bird bath and feeder will give them somewhere to wash and food to eat. For the perfect sugary treat, fill your feeder with four-parts water, one-part sugar. Don't forget to add a tiny perch for them to rest on.

3. Hummingbirds also snack on insects and spiders. Hang a basket of banana peels from a branch to attract small bugs and you may attract hummingbirds, too!

Spotting Hummingbirds

WHERE TO SEE THEM

In backyards, rooftop gardens, balconies, window boxes, parks and botanical gardens.

WHEN TO SEE THEM

In the morning or late afternoon whizzing from flower to flower.

LISTEN OUT FOR...

The humming of their wings or loud chirps when they're chasing each other!

TURKEY VULTURES
in Havana, Cuba

ONCE THE HAUNT OF PIRATES AND REVOLUTIONARIES, HAVANA IS A COLOURFUL CITY WITH GRAND OLD BUILDINGS AND COBBLED PLAZAS. Look up and you might see some of nature's most feared and misunderstood urban birds circling ominously in the sky—turkey vultures! Some people think vultures are ugly scavengers that swoop in to pick over the scraps of dead animals, but there's much more to these clever creatures than meets the eye...

Cities are warmer than the vulture's natural habitat. To cool down, they cover their legs with their own urine and droppings! Their droppings are also very acidic, which helps to kill off harmful bacteria from rotting carcasses.

Powerful hooked beak for tearing up food

Featherless head that makes it easier for it to clean itself after poking its head into animal carcasses

Long toes with blunted talons, which are easier for walking

LATIN NAME *Cathartes aura*
FAMILY Cathartidae (New World vultures and condors)
LENGTH 62–81 cm (2–2.7 ft)
WINGSPAN 1.6–1.8 m (5.2–5.9 ft)
CONSERVATION STATUS Least Concern
WHERE IN THE WORLD? North and South America

City Life

Vultures can be seen in many cities around the world. They use buildings as lookout posts or as roosts (places to sleep) and lay their eggs on high ledges or in abandoned buildings. And they're not too picky about what they eat. They walk the streets feeding on food waste and small animals that have been killed by cars and trucks.

In Havana, turkey vultures can often be seen circling the famous Jose Marti Memorial in Revolution Square.

FEATHERED SUNBATHERS

Each morning, vultures stretch their wings out wide to warm and dry their feathers in the sun before heading out for the day. In cities, they can be seen sunning themselves on rooftops, fence posts, telegraph poles, or on top of cars!

Long broad wings with 'fingers' at the tips — perfect for soaring long distances without using up too much energy

Havana is famous for its old-fashioned cars that date from the time of the Cuban revolution in the 1950s.

The Native American Cherokee people know the turkey vulture as the 'peace vulture', because it doesn't usually kill other creatures. It feeds mainly on dead animals, known as carrion.

CLEANING THE CITY

They may not be very pretty, but vultures can be useful to have around. They're part of nature's clean-up crew, clearing cities of dead animals and helping to stop the spread of dangerous diseases.

Turkey vultures have an amazing sense of smell. They can sniff out rotten food more than a mile away.

In the Brazilian city of Belém, black vultures crowd around rubbish dumps and help fishermen to clear away leftover fish carcasses.

Spotting Vultures

They can often be spotted near airports where lots of small animals are killed by trucks and aircraft.

Sometimes they fly too close and crash into plane windscreens or engines.

WHERE TO SEE THEM

Huddled around rubbish bins or roadkill (an animal that has been killed on the road by a vehicle). Gliding low to the ground sniffing for food or circling in the sky.

WHEN TO SEE THEM

In the morning as they sun their wings.

LISTEN OUT FOR . . .

Vultures don't have a syrinx (the vocal organ that most birds have) so they can't sing. Instead, they make soft hissing, sneezing and grunting noises.

MORE ANIMALS

KINGS OF THE CITY

Rats have taken to city life like fish to water. They can squeeze through a hole the size of a pen and will set up home in derelict buildings, under floorboards, inside walls and in drainpipes, alleyways and sewers. They'll eat pretty much anything, too, from nuts and seeds to scrambled eggs and hamburgers. The stomach contents of one rat revealed over 4000 items!

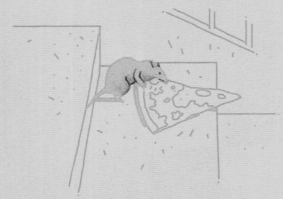

One city rat was spotted carrying a piece of pizza down the stairs of a New York City subway.

Their super-sharp teeth can nibble through wood, electrical cables and even cement!

Rats are amazing swimmers and can tread water for up to three days. They love hanging out in city sewers and sometimes swim up through toilet pipes.

RATS OF THE SKY?

Some see them as 'rats of the sky' that splatter statues with their droppings. Others like having them around and even feed them. Love them or hate them, pigeons are everywhere! Rooftops and ledges are a perfect substitute for rocky sea cliffs, and they're happy snacking on whatever we throw away. They are also skilled flyers and can dodge between buildings and through moving traffic.

These super-adaptable birds can be found pecking crumbs from pavements in cities all around the world.

NUT-CRACKING CROWS

In the Japanese city of Akita, crows have developed an ingenious way of cracking open nuts. They position themselves by pedestrian crossings, drop the nut into the path of oncoming traffic and let the cars do the rest. Next, they wait for the lights to turn red, then they hop onto the road to collect their dinner.

IN THE CITY

Here are some more city-slicker creatures and the surprising ways they've adapted to the hustle and bustle of city life.

LEAPING LIZARDS

Small tropical lizards called anoles living in the Puerto Rican city of San Juan are thought to have longer legs with more grip than those in forests. This makes it easier for them to scuttle up smooth metal and glass surfaces. The common agama lizard living in Monrovia, Liberia, has another trick up its sleeve. It can communicate by dancing, so it doesn't have to shout to be heard over the noise of the city!

FREE FISH SUPPER

On Santa Cruz Island in the Galápagos, Ecuador, sea lions have learned to stand in line at a local fish stall — just like the other customers. They share the queue with pelicans, gulls and the occasional marine iguana.

One sea lion stood on its flippers for an hour while it patiently waited its turn!

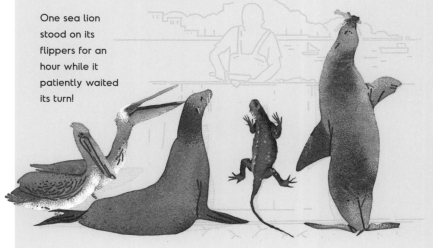

CITY SINGERS

Scientists have discovered that some city birds sing faster and higher-pitched songs than those living in forests, who sing lower and slower.

Great tits recorded near Buckingham Palace in London, UK, sing fast and high so they can be heard above the noisy traffic.

HIPPO LAWN MOWERS

The last thing you'd expect to see in your back garden is a hippopotamus nibbling your lawn...unless you live in the sleepy town of St Lucia, South Africa. As night falls, the locals head indoors while hippos from the nearby iSimangaliso Wetland Park stroll the streets to graze on grass.

WHAT YOU CAN

AS OUR CITIES CONTINUE TO GROW, people are finding new ways to make them more wildlife-friendly — from creating rooftop lakes and gardens to building under-road tunnels for frogs and koalas, or rope bridges for possums. Here are some tips on what you can do to help.

TIP 1. POLLINATOR PIT STOPS

Help save bees, butterflies and other pollinators by making sure they have lots of places to refuel throughout the city.

FLOWER POWER: Pollinators feed on the nectar of flowering plants. Find out the favourite flowers of your local bees and butterflies then plant some in your garden. If you don't have a garden, you can grow plants in a hanging basket or a pot. Every flower counts!

TOP TIP: An old mushy banana will give butterflies a sugary treat!

TIP 2. BACKYARD JUNGLES

Don't be too neat and tidy. Piles of old leaves, or weeds that grow on lawns, by the road, or in between the cracks in walls, can mean food and shelter for all kinds of animals.

Many caterpillars love to eat garden weeds such as dandelions, brambles, nettles and grasses.

A stack of rocks, sticks and old flowerpots filled with hollow stems can make the perfect hideaway for bugs and lizards.

Pile up dead leaves in a sheltered spot to give small mammals somewhere warm to hide, sleep and hibernate.

TIP 3. COSY COMPOSTS

Compost bins are a cosy resting spot for reptiles and amphibians on cold days. They're also popular with worms, which help break down the waste and turn it into soil you can use on your garden.

All you need is a compost bin or tub. Add vegetable peelings, old teabags and grass cuttings. Then watch to see who moves in.

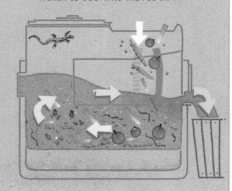

TIP 4. FEEDERS AND NEST BOXES

Feeders and nest boxes provide food and shelter for all kinds of urban birds from sparrows and swifts to pheasants and ducks.

If you're trying to attract birds to your garden, you should keep your cat indoors or put a bell on the cat's collar so birds can hear it coming.

Do to Help

WATER STOP: Leave a shallow dish of water nearby so pollinators can have a drink if they want to. You can also add some pebbles for bees to sit on.

SUN LOUNGERS: Butterflies need sunshine to give them energy and to help them fly. Put some small rocks in a warm spot to soak up the rays and give butterflies somewhere to sunbathe.

SHADY HIDEAWAY: Flying from flower to flower can be exhausting, especially on sunny days. Give butterflies and bees a cool place to rest by growing leafy plants with lots of shady cover.

TIP 6. HEDGEHOG HIGHWAYS

Wooden fences can make it difficult for small mammals such as hedgehogs to move from garden to garden. Cut a small doorway at the bottom of the fence to create your very own hedgehog highway.

HEDGEHOGS THIS WAY

TIP 5. PONDS

A garden pond can be a great place for frogs, toads and dragonflies to lay their eggs and grow up. It doesn't need to be big — you can even use an old washing-up bowl to make a mini pond. Add pond plants and rocks to give animals a place to hide, and a small ramp to help animals crawl in and out.

TIP 7. KEEP IT NATURAL

Pesticides may help get rid of garden pests, but they can be harmful to other bugs and bigger animals, too. By letting the good bugs, birds and lizards eat unwanted critters you can keep your garden in balance.

GLOSSARY

adapt To change your behaviour to suit new conditions (e.g. to make it easier to live in a particular place)

adaptation A special skill or characteristic that helps an animal to survive

aerodynamic The more aerodynamic an object is, the better it will fly

amphibian A cold-blooded, semiaquatic animal with a backbone (e.g. a frog, toad, salamander or caecilian). Most amphibians have gills and live in the water when they are young and breathe air and live on land as an adult. The word amphibian means 'two lives'

antenna (*plural*: **antennae**) A long, thin movable body part found on the heads of insects and crustaceans, which they use to feel and smell things around them

beak The hard, pointed part of a bird's mouth used for eating, preening, fighting and courting

bill Another word for a beak

biodiverse A habitat or region where there are lots of different species of plants and animals

caecilian A limbless amphibian that looks like a large worm

camouflage The way in which some animals are coloured or shaped to help them blend in with their surroundings

carnivore An animal that eats meat

carcass The body of a dead animal

carrion The dead or rotting body of an animal

chrysalis A butterfly or moth in the stage between being a larva and a fully grown adult when it grows a protective covering and stops moving and eating

cold-blooded Cold-blooded animals, such as reptiles, amphibians and fish, rely on heat from their surroundings to warm them up

colony A group of animals of the same type that live together

communal Something that is shared by a group of people or animals, such as a home (e.g. some species of birds live together in communal nests)

conservation Nature conservation is the protection of the natural world including all animals and plants

crest A tuft on the head of an animal

diurnal Diurnal animals are active during the day

endangered An endangered species is one that is in danger of becoming extinct (disappearing forever)

endemic Found only in one particular place. If an animal or plant is endemic to a country or region it means it only lives there

forage To search for food

gill The organ through which fish and other water creatures breathe

habitat The natural home of an animal, plant or other organism (living thing)

insect A small invertebrate animal that has six legs

invertebrate An animal that doesn't have a backbone, for example an insect, a spider, a worm or an octopus

larva (*plural*: **larvae**) The young form of an insect or an animal such as a frog that looks very different to its adult form

mammal A warm-blooded animal that breathes air, has a backbone and grows hair or fur. Most female mammals give birth to live babies rather than laying eggs and feed their young milk

megacity A large or important city

metamorphosis A process some animals go through to become adults in which they change from one form to a very different form

metropolis A large or important city

migration Some animals move from one place to another at certain times of the year. This is called migration

nectar A very sweet, sugary liquid made by the flowers of plants to attract pollinating animals

nocturnal Nocturnal animals are active at night

omnivore An animal that eats both meat and plants

pagoda A tower common in parts of Asia, often built for religious purposes

pesticide A substance used to destroy pests such as insects and weeds

plaza A public square in a city or town

plume A feather of a bird

pod A small group of whales or dolphins

pollen Very tiny, dust-like grains produced by flowers that cause plants to grow seeds. New plants can then grow from the seeds. Pollen is carried from one plant to another by the wind or by animals such as bees and hummingbirds

pollination When pollen is transferred from one plant to another

pollinator An animal that carries pollen from plant to plant, helping to create seeds

predator An animal that kills and eats other animals

prehensile Adapted for grasping and holding onto things by wrapping around it (e.g. some monkeys have a prehensile tail)

prey An animal that is hunted and eaten by another animal

pupa (*plural*: **pupae**) An insect in the stage between being a larva and a fully grown adult. It has a protective covering and doesn't usually move (e.g. a chrysalis)

reptile A cold-blooded, air-breathing animal, with a backbone and with scales instead of hair or feathers

roost A place where birds or bats rest or sleep

scavenger An animal that feeds on plants, rubbish or carrion (dead animals)

souk A traditional marketplace in northern Africa or the Middle East

suburb The edge of a town or city where there are mainly houses, schools and shops and fewer big buildings

thermal A rising current of warm air. Birds use thermals to help lift them through the air

tusk A long, pointed tooth for fighting or digging

vegetation The plant life of a region

venom A toxic substance produced by some animals (such as snakes) that is injected into their prey

vertebrate An animal that has a backbone

warm-blooded Warm-blooded animals (mammals and birds) can make their own body heat even when it is very cold outside

wetlands Areas of land where the ground is very wet (such as bogs, swamps and marshes). Wetlands can be home to a large number of plants and animals

INDEX

ACKNOWLEDGEMENTS

Published in November 2019
by Lonely Planet Global Limited
CRN 554153
ISBN 978 1 78868 490 3
www.lonelyplanetkids.com
© Lonely Planet 2019
10 9 8 7 6 5 4 3 2 1
Printed in China

Publishing Director: Piers Pickard
Publisher: Hanna Otero
Concept and Design: Tina García
Art Director: Andy Mansfield
Author and Editor: Kate Baker
Illustrator: Gianluca Folì
Consultant: Dr Ricardo Rocha
Print Production: Lisa Ford

Special thanks go to the consultant Dr Ricardo Rocha, Researcher in Conservation Science at the University of Cambridge, UK, for all his help in putting this book together. Thanks also to Dr Leonardo Ancillotto, Researcher in Wildlife Ecology at University of Naples Federico II, Italy; Joshua I. Brian, MSc, Researcher in Conservation Science at the University of Cambridge, UK; Albano Soares, Researcher in Entomology at TAGIS — Centro de Conservação das Borboletas de Portugal; and Dr Piero Visconti, Researcher in Global Change Biology at the International Institute for Applied System Analysis, Austria.

STAY IN TOUCH lonelyplanet.com/contact

AUSTRALIA
The Malt Store, Level 3, 551 Swanston St,
Carlton, Victoria 3053 T: 03 8379 8000

USA
124 Linden St, Oakland, CA 94607
T: 510 250 6400

IRELAND
Digital Depot, Roe Lane (off Thomas St), Digital Hub,
Dublin 8, D08 TCV4

UNITED KINGDOM
240 Blackfriars Rd, London SE1 8NW
T: 020 3771 5100